GENERAL
PHOTOBIOLOGY

Titles of related interest

FAHN	PLANT ANATOMY, 3RD EDITION
GOODWIN & MERCER	PLANT BIOCHEMISTRY, 2ND EDITION
OSBORNE	PROGRESS IN RETINAL RESEARCH

JOURNALS

Phytochemistry & Photobiology

Vision Research

GENERAL
PHOTOBIOLOGY

by

DONAT-PETER HÄDER

and

MANFRED TEVINI

PERGAMON PRESS

OXFORD · NEW YORK · BEIJING · FRANKFURT
SÃO PAULO · SYDNEY · TOKYO · TORONTO

U.K.	Pergamon Press, Headington Hill Hall, Oxford OX3 0BW, England
U.S.A.	Pergamon Press, Maxwell House, Fairview Park, Elmsford, New York 10523, U.S.A.
PEOPLE'S REPUBLIC OF CHINA	Pergamon Press, Qianmen Hotel, Beijing, People's Republic of China
FEDERAL REPUBLIC OF GERMANY	Pergamon Press, Hammerweg 6, D-6242 Kronberg, Federal Republic of Germany
BRAZIL	Pergamon Editora, Rua Eça de Queiros, 346, CEP 04011, São Paulo, Brazil
AUSTRALIA	Pergamon Press Australia, P.O. Box 544, Potts Point, N.S.W. 2011, Australia
JAPAN	Pergamon Press, 8th Floor, Matsuoka Central Building, 1-7-1 Nishishinjuku, Shinjuku-ku, Tokyo 160, Japan
CANADA	Pergamon Press Canada, Suite 104, 150 Consumers Road, Willowdale, Ontario M2J 1P9, Canada

First edition 1987

Library of Congress Cataloging in Publication Data
Häder, Donat-Peter.
General photobiology.
(Pergamon international library of science, technology, engineering, and social studies)
Bibliography: p.
1. Photobiology.   I. Tevini, Manfred, 1939–
II. Title.   III. Series.
QH515.H33 1987   574.19′153   86–21222

British Library Cataloguing in Publication Data
Häder, Donat-Peter.
General photobiology.
1. Photobiology
I. Title   II. Tevini, Manfred
574.19′153   QH515

ISBN 0–08–032027–9 Hardcover
ISBN 0–08–032028–7 Flexicover

Printed in Great Britain by A. Wheaton & Co. Ltd., Exeter

Preface

Photobiology is an interdisciplinary science which during the past few years has undergone a dramatic development. Despite its central role in biology, medicine, physics and chemistry, until now no modern introductory text has been available. Photobiology is rooted in quite diverse fields of science which makes it difficult for the beginner to comprehend. Therefore we have decided to introduce the photochemical and photophysical basics in the opening chapters rather than to assume a solid knowledge of the required background.

The first chapter describes the fundamental photophysics with a special emphasis on the terminology, using the internationally accepted Si system of units. Studying the current photobiological literature shows that many scientists have problems using the physical units in the energetic and visual systems. Therefore we include a table with the most important physical units. Since this book is not intended to present the theoretical background, but to supply the experimenter with practical clues, the following chapters describe the commonly used light sources, filters and monochromators, as well as instruments and methods to measure radiation. The emphasis is on practical usage, and therefore we discuss the limitations of the techniques and the possible sources of error, without assuming an intimate mathematical knowledge.

Teaching photobiology in class reveals that students without a solid background in chemistry have difficulties understanding the photochemical processes at the level of electrons, atoms and molecules. Therefore the introductory part contains a detailed description of the basic photochemical reactions and the energy transfer during excitation and relaxation of molecules.

The main part of the book covers the classical photochemical topics subdivided into processes in which energy is derived from light, such as photosynthesis, and those in which light is used as a sensory signal. This second group comprises the effects of light on development (photomorphogenesis), on orientation of plants and microorganisms (photomovement) and on sensory systems of animals and men. The final chapter describes the damaging effects of UV and the possible repair mechanisms.

This book is not intended to be an encyclopedic collection of reviews but rather attempts to present the subject using relevant examples. The selection is naturally subjective. The most important goal is always to explain the underlying general principle.

The wide scope of this interdisciplinary subject prohibits any individual to completely master the whole field; therefore we have relied on critical and competent help from our colleagues, to whom we are indebted. We appreciate the constructive criticism and productive discussions with Dr Braslavsky, Prof. Hartmann, Prof. Haupt, Priv.-Doz. Jabben, Prof. Nultsch, Dr Pfister, Prof. Scheer, Prof. Stieve, Prof. Trebst and Prof. Wagner. We would like to thank Prof. Song, who critically read the translated version. We are also very grateful to Dipl. Biol. G. Traxler for her help with the translation and to Mr Bosch, who completed most of the line drawings.

Marburg and Karlsruhe   D.-P. Häder
Spring 1986   M. Tevini

Contents

Important Units in the SI System

Basic units

length	meter	[m]
mass	kilogram	[kg]
time	second	[s]
electric current	ampere	[A]
temperature	kelvin	[K]
luminous intensity	candela	[cd]
quantity	mol	[mol]

Derived units

frequency	Hertz	$[Hz] = [s^{-1}]$
force	Newton	$[N] = [kg\ m\ s^{-2}]$
energy	Joule	$[J] = [W\ s] = [kg\ m^2\ s^{-2}]$
power	Watt	$[W] = [kg\ m^2\ s^{-3}]$
pressure	Pascal	$[Pa] = [N\ m^{-2}] = [kg\ m^{-1}\ s^{-2}]$
electric potential	Volt	$[V] = [J\ A^{-1}\ s^{-1}] = [W\ A^{-1}]$
electric charge	Coulomb	$[C] = [A\ s] = [J\ V^{-1}]$
magnetic flux	Weber	$[Wb] = [W\ s\ A^{-1}]$
magnetic induction	Henry	$[H] = [V\ s\ A^{-1}]$

Photophysical units

(a) Based on photons

quantity of photons	[mol]
steric photon density	$[mol\ sr^{-1}]$
photon fluence	$[mol\ m^{-2}]$
photon flow	$[mol\ s^{-1}]$
photon fluence rate	$[mol\ s^{-1}\ m^{-2}]$

(b) Based on energy

radiant energy	$[J] = [W\ s]$
steric energy flux	$[J\ sr^{-1}] = [W\ s\ sr^{-1}]$

energy fluence $[\text{J m}^{-2}]$
energy flow $[\text{W}] = [\text{J s}^{-1}]$
energy fluence rate ($=$ flux) $[\text{W m}^{-2}] = [\text{J s}^{-1} \text{m}^{-2}]$

(c) *Photometrical units*
luminous intensity $[\text{cd}]$
luminous flux $[\text{lm}] = [\text{cd sr}]$
luminous energy (light quantity) $[\text{talbot}] = [\text{lm s}]$
luminance $[\text{nit}] = [\text{cd m}^{-2}]$
illumination $=$ light flux $[\text{lx}] = [\text{lm m}^{-2}] = [\text{cd sr m}^{-2}]$
photometric fluence $[\text{lx s}] = [\text{lm s m}^{-2}]$

Decimal prefixes

tera- (T) 10^{12} centi- (c) 10^{-2}
giga- (G) 10^{9} milli- (m) 10^{-3}
mega- (M) 10^{6} micro- (μ) 10^{-6}
kilo- (k) 10^{3} nano- (n) 10^{-9}
hecto- (h) 10^{2} pico- (p) 10^{-12}
deca- (da) 10^{1} femto- (f) 10^{-15}
deci- (d) 10^{-1} atto- (a) 10^{-18}

Fundamental constants

Gas constant $8.314 \text{ J mol}^{-1} \text{ K}^{-1}$
Avogadro's number $6.022 \times 10^{23} \text{mol}^{-1}$
Boltzmann's constant $1.381 \times 10^{-23} \text{ J K}^{-1}$
Elementary electric charge $1.602 \times 10^{-19} \text{ C}$
Gravitational acceleration 9.806 m s^{-2}
Faraday's constant $9.649 \times 10^{4} \text{ C mol}^{-2}$
Velocity of light $2.998 \times 10^{8} \text{ m s}^{-1}$
Planck's constant $6.626 \times 10^{-34} \text{ J s}$

Introduction

... the sun is not only the author of visibility in all visible things, but of generation and nourishment and growth ...

Plato, *The Republic*

The radiation from the sun is the single most important source of energy for life on our planet. The sun supplies the total energy for the atmospheric and climatic phenomena which cause the physical and chemical changes on the earth's surface. In addition, solar radiation has played a key role during the evolution of organisms on earth. The solar radiation is the result of nuclear fusion and covers the wavelength range from about 200 to 3200 nm when measured outside our atmosphere. The basic unit for all calculations is the solar constant, which is the extraterrestrial radiant energy which passes through a unit area perpendicular to the incident rays during 1 minute (measured at the mean solar distance). In the stated wavelength range, the solar constant amounts to about 1350 W m^{-2}. Due to absorption by oxygen, ozone, carbon dioxide and water vapour in the atmosphere the global radiation which reaches the earth's surface is about 17% less than the solar constant. Ultraviolet radiation below 300 nm is selectively filtered out of the solar spectrum by ozone, which evolves from oxygen in a photochemical reaction in the stratosphere.

The evolution of life on this planet began at a time when no oxidizing atmosphere existed which allowed the short ultraviolet radiation to reach the surface. Among other factors, it was this radiation which induced chemical evolution and caused the important changes in the already existing molecules, such as the nucleic acids. In their experiments Miller and Ponnamperuma demonstrated the photochemical production of organic molecules such as amino acids in the presence of the reducing atmosphere which is believed to have covered the earth during the initial phase of evolution.

The effect of high-energy radiation on the genesis of self-reproducing nucleic acids as the basis for the reproduction of organisms is still obscure. After the chemical evolution of light-absorbing pigments such as chlorophylls and carotenoids, bacteria-like organisms succeeded in harvesting the solar energy and transforming it into chemical energy. This process, called photosynthesis, results in the production of high-energy organic compounds (biomass) from low-energy precursors. This development gave the phototrophic organisms an ecological advantage over the chemosynthetic organisms, since solar energy is available in a virtually unlimited supply. Fossil fuels such as oil and coal result from the photosynthetic biomass production of higher and lower plants. The first primitive photosynthetic pathways operated without oxygen production. Photosynthetic bacteria still exist as representatives for this anoxigenic photosynthesis. The next step in evolution was the development of a water-splitting system which generated the oxygen in our atmosphere. Evolution could then proceed sheltered from the damaging short-wavelength UV radiation by the stratospheric ozone layer.

In addition to energy fixation, microorganisms, plants and animals use light for a wide range of complex sensory processes. It is not only aminals that orient in light and search actively for a favorable range of light intensities in their environment. Microorganisms use light-dependent movement responses to find a suitable niche in their photoenvironment. Even rooted higher plants orient with respect to light, as can easily be observed in pot plants which turn their leaves toward the window. The necessity for photoorientation is obvious in photosynthetic organisms which depend on the availability of radiation, but non-photosynthetic organisms also use photoorientation, for example to escape extreme light intensities which might bleach their pigments and damage cellular components.

In addition, light triggers developmental rhythms in plants and animals. It would be awkward if birds used the ambient temperature as a signal to initiate their annual migrations, since a cool late summer or a warm autumn would disturb their time schedule. Instead they use the day length as a very precise timer. Likewise plants utilize this parameter as a signal for flower initiation or leaf abscission. Both plants and animals are capable of measuring the day length very exactly and compare it with their internal oscillator ('*Zeitgeber*'). The morphology of an organism is controlled by light. When a plantlet grows out of the soil it develops leaves only above the surface. Light induces leaf growth and stimulates chloroplast development since in darkness the photosynthetic apparatus would be useless.

The basic photochemical and photophysical reactions are the same in all organisms regardless of whether the absorbed radiation is used for energy fixation or to control development or movement. A quantum of light of a defined wavelength is absorbed by a photoreceptor molecule. This process excites an electron which undergoes a transition to an energetically higher state. This energy can be lost again when the molecule returns to its ground state by emission of a quantum or by losing the energy as heat. As an alternative the energy of the excited state can be utilized for chemical reactions.

The book is divided into three parts. First we introduce the photophysical and photochemical principles of excitation and relaxation of photoreceptor molecules. Then we discuss the energy-fixing processes in bacteria and higher plants. Finally we cover the sensory processes in microorganisms, plants, animals and men, controlled by light.

PART I

General Photophysics

1   Fundamental photophysics

1.1   THE ELECTROMAGNETIC SPECTRUM

Light is defined as a small range in the spectrum of the electromagnetic waves. These waves can be characterized either by their frequency (number of oscillations per second = Hertz, indicated by the Greek letter v) or by their wavelength (in m, indicated by the Greek letter λ) which can easily be transformed into each other using the following equation:

$$v = \frac{c}{\lambda}.$$

The velocity of electromagnetic waves in a vacuum (c) is 2.9979×10^8 m s^{-1}. The frequency of electromagnetic waves covers the range from a few Hertz to infinity (Fig. 1.1). Technical alternating currents have frequencies up to about 200 Hz, followed by audio frequencies (about 20 Hz–20 kHz) which are important for communication technologies. The next frequency band covers the radio frequencies. Television and radar utilize even higher frequencies. Oscillations with wavelengths in the centimeter and millimeter range can be produced by electronic circuits while oscillations with shorter wavelengths (higher frequencies) are exclusively generated by atoms and molecules. For biological processes the band between 200 and 900 nm (about 10^{15} Hz)[1] is especially important. In this spectral range we find the visible radiation seen by humans as light of different colors. In addition to the visible radiation, the adjacent bands with longer wavelengths (infrared, IR) and shorter wavelengths (ultraviolet, UV) are also of biological interest. This radiation can be used both for energetic and for regulatory purposes and it can have a damaging effect. In optics we usually deal with wavelengths while in electronics, infrared and Raman spectroscopy people work with frequencies or wave numbers [cm^{-1}].

The short wavelengths in the visible range are seen as purple followed by blue, green, yellow and orange. At the long wavelength end the visible spectrum is limited by red and far red. A superimposition of all these

---

[1] DIN 5031, section 3 defines the visible radiation as the wavelength range between 380 nm and 780 nm where the spectral sensitivity $V(\lambda)$ (cf. section 1.4.2) is 0.001% of the maximal sensitivity.

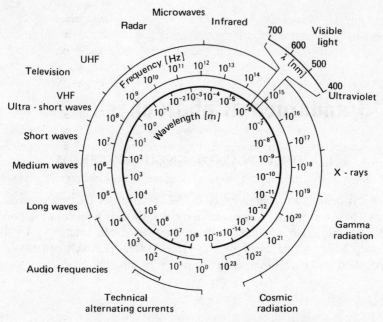

Fig. 1.1. Spectrum of the electromagnetic waves. Inner scale: wavelengths in meters; central scale: frequencies in Hertz; outer scale: name and usage of waves. The enlarged inset shows the range of visible light.

wavelengths appears as white light. Ultraviolet, invisible to the human eye, can be subdivided according to its different effects on biological systems (see Chapter 2): UV-A defines the range between 320 nm and 400 nm. The band between 280 nm and 320 nm is called UV-B, followed by the UV-C band with even shorter wavelengths. The higher frequency range starting at about 10^{16} Hz comprises the X-rays, gamma and cosmic rays.

1.2   LIGHT RAYS AS WAVES

In 1690 Huygens published his wave theory in a paper called 'Traité de la lumière'. Later his theory was further developed by Fresnel who regarded light as a transversal wave in an elastic medium called 'ether'. Today we know that longitudinal mechanical waves as well as sound waves require the existence of matter (due to the lack of an atmosphere on the moon you do not hear what someone else says), while light does not require the presence of matter. Of course, the interstellar space is not free of matter but electromagnetic waves can also travel in a vacuum.

According to Maxwell's theory, travelling electromagnetic waves result from a rapid alternation of electric and magnetic fields which induce each other and are oriented perpendicular to each other. Electromagnetic fields are generated by vibrations of atoms or molecules, which do not produce continuous waves but rather a train of several oscillations. Since waves originating from different molecules have vibration planes not parallel to each other, radiation from the sun or a lamp is a mixture of differently polarized waves. In addition, the waves differ in their phases. Such radiation is defined as not being coherent; coherent waves originate from sources with the same frequency and have the same phase. Radiation of one distinct wavelength is called monochromatic but this term is used in the laboratory in a broader sense describing radiation in a narrow wavelength range such as 10 nm.

1.3 QUANTUM CHARACTERISTICS OF LIGHT

Many characteristics of light, such as diffraction, refraction, interference and its travelling in space, can be described by assuming light to be a wave. When we study the electrical response of a photomultiplier (see Section 4.1.1) using very low fluence rates we observe a sequence of single pulses which can be explained only by assuming the absorption of distinct particles (photons). In 1675, Newton proposed a theory according to which radiation consists of small particles which are emitted by a source and which can travel through transparent materials. Modern physics assumes both wave and quantum characteristics of electromagnetic radiation. This dualistic radiation theory was initiated by Planck (1900), Einstein (1905) and de Broglie (1924).

Since photons, in contrast to hadrons (neutrons and protons) or electrons, have no resting mass, we call them virtual particles. However, photons have a defined energy which, as Planck proved, depends on the frequency and thus on the wavelength of the radiation

$$E = h\nu = h\frac{c}{\lambda}$$

where h, Planck's constant, is 6.63×10^{-34} W s^2.

In the following example we calculate the energy of a quantum of light with a wavelength of 500 nm

$$E = \frac{6.63 \times 10^{-34} \text{ W s}^2 \times 2.9979 \times 10^8 \text{ m s}^{-1}}{5 \times 10^{-7} \text{ m}} = 3.98 \times 10^{-19} \text{ W s.}$$

Since 1 W s equals 6.242×10^{18} eV the energy of this quantum corresponds to 2.48 eV. One mole of quanta (6.02×10^{23}) is defined as 1 Einstein and has an energy (provided the wavelength is 500 nm) of:

$$6.02 \times 10^{23} \times 3.97 \times 10^{-19} \text{ W s} = 2.39 \times 10^5 \text{ W s.}$$

Since the energy of a quantum is inversely related to its wavelength we should stress the important fact that radiation with shorter wavelengths has a higher energy than that with longer wavelengths, which will be important for the following discussions. In fact, purple photons (400 nm) carry almost double the energy than that carried by dark red photons (700 nm).

1.4   THE PHYSICAL PROPERTIES OF LIGHT

Since 1 January 1978 the usage of the SI units (Système International d'Unités) has been obligatory for the calculation of physical units. The meter [m] is the basic unit of length, kilogram [kg] of mass, second [s] of time, ampere [A] of electric current, Kelvin [K] of thermodynamic temperature, candela [cd] of luminous intensity and mole [mol] of the amount of a substance. In addition, there are a number of derived units.

In the following we will define the photophysical terms with their units in the SI system. While the physical units are unambiguous, photobiologists, physicists and light engineers use a number of different terms which we discuss comparatively.

1.4.1   Radiant energetic units

We assume a point source which emits a certain number of particles (photons). This photon number can be described by a dimensionless number (N) with the unit [mol]. Since, as we have seen above, each photon has a certain energy (corresponding to its wavelength) we can describe the radiation by its radiant energy calculated in Joules [J] or [W s].

To calculate the radiant energy emitted into a certain sector, we are interested in the solid angle (ω). In a plane we define an angle either in degrees or radians. 1 radian is the angle enclosed by two lines which cut out an arc from the circumference which equals the radius r. Since the circumference of $2\pi r$ corresponds to 360°, 1 radian amounts to about 57.3°. When we use the same calculation in three dimensions we define a solid angle of 1 steradian [sr] as the angle of a cone which subtends an area $F = r^2$ within the surface of a sphere with the radius r.

Now we can define a steric particle flux F as the number of photons per solid angle with the unit [mol sr^{-1}] and the radiated energy in a sector as steric energy flux M with the dimension [J sr^{-1}]. Since in our example of a point source photons are emitted in all directions of space, the number of photons per unit area drops with the distance squared.

Now we reverse the point of view and focus out attention to the absorbing object, especially since we are interested in the impinging (or to be more accurate: absorbed) quanta which produce a photochemical effect. To facilitate the calculations we consider a beam of parallel rays impinging perpendicular to the target. We define a photon exposure Y as the number of photons impinging on a given surface with the unit [mol m^{-2}] and radiant exposure Ψ (=energy area density) as the energy of the photons per unit area with the unit [J m^{-2}].

In these considerations time does not play a role, as we can deduce from this simple example. When we produce a photographic positive from a negative we need a certain number of photons per unit area to produce a certain gray level. Within limits it is not important if the necessary number of photons is applied within a short period of time (wide open aperture) or a long period of time (almost closed aperture). The energy absorbed by a unit area during the exposure time is also called a dose [J m^{-2}].

We can also ask for the number of photons emitted per time unit which defines the photon flow A with the unit [mol s^{-1}] which corresponds to a radiant flow (=energy flux) Φ which describes the emitted energy per time unit ([J s^{-1}]=[W]) and defines a unit of power. When we follow the energy flux from a point source a fraction I is emitted into a certain sector [J s^{-1} sr^{-1}].

The biologist is more interested to know the radiant flux or the number of photons per time unit impinging on an object such as a leaf. For the moment we neglect the fact that only a certain percentage of the photons is absorbed. The photon irradiance Γ [mol s^{-1} m^{-2}] corresponds with a radiant flux density E [J m^{-2}]=[W m^{-2}], which is also called energy fluence rate, energy flosan or (energy) irradiance depending on the mode of irradiation.

As stated above, the energy of a quantum depends on the wavelength of the radiation; therefore we use the subscript λ to indicate this dependence: Q_λ, M_λ, Ψ_λ, Φ_λ, E_λ, I_λ, and define a spectral radiant energy, spectral radiant flux density etc. (Table 1.1).

In the older literature especially, we find a number of units which are no longer permitted in the SI system. In order to facilitate the calculation the following example shows the conversion of units from [erg cm^{-2} s^{-1}]

$$1000 \text{ erg cm}^{-2} \text{ s}^{-1} = 1 \text{ J s}^{-1} \text{ m}^{-2} = 1 \text{ W m}^{-2}$$

Smaller radiant flux densities should be described using prefixes [mW m^{-2}] or [μW m^{-2}] instead of using smaller areas [cm^2].

$$1 \text{ W m}^{-2} = 1000 \text{ mW m}^{-2} = 10^6 \mu\text{W m}^{-2} = 100 \ \mu\text{W cm}^{-2}$$

TABLE 1.1   UNITS AND DIMENSIONS IN THE ENERGETIC SYSTEM

Quanta			Based on		
	Solid angle	Area	Time	Time and area	Solid angle and time
N [mol]	F [mol sr^{-1}]	Y [mol m^{-2}]	A [mol s^{-1}]	T [mol s^{-1} m^{-2}]	[mol s^{-1} sr^{-1}]
Energy Q_e [J] = [W s]	M_e [J sr^{-1}]	ψ_e [J m^{-2}]	Φ_e [J s^{-1}] = [W]	E_e [J s^{-1} m^{-2}] = [W m^{-2}]	I_e [J s^{-1} sr^{-1}] = [W sr^{-1}]

All known biological effects of radiation on organisms depend on the number of absorbed quanta rather than on the incident energy. Therefore it is useful to indicate the number of quanta. One quantum at a wavelength of 666 nm has an energy of 2.977×10^{-19} W s. One mole of quanta of this wavelength thus has an energy of 1.792×10^5 W s. Using this relationship, we calculate that 1 W m^{-2} corresponds with 5.578 μW m^{-2} s^{-1} for a wavelength of 666 nm.

All these previous considerations have assumed a parallel and normal irradiation, which is a special case found under many experimental conditions. Natural conditions, however, are far more complicated. Multiple scattering and reflection change the optical path in such a way that the irradiation impinges from all directions on to an object such as a plant. A planar photoreceptor (e.g. a leaf) is thus irradiated from a hemispherical space. The total irradiation is calculated by integrating over this hemisphere (cosine correction). Under natural conditions it is important to distinguish between normal and parallel irradiation, and radiation from a hemispherical space or from all directions.

1.4.2   Photometric units

The previous section concentrated on the quantity and energy of photons; therefore all symbols should carry the index 'e' to distinguish them from the photometric symbols. The photometric system of units is based on the specific absorption characteristics of the human eye. The definitions use the visual impression a given radiation induces in our eye; therefore, all photometric symbols carry the index 'v' (Table 1.2). The spectral sensitivity of the human eye is shown in Fig. 1.2 on a linear and a logarithmic scale. The maximal sensitivity of the bright-adapted eye is at 555 nm. At wavelengths of 470 nm and 650 nm the sensitivity is only about 10% of that of the maximum. A

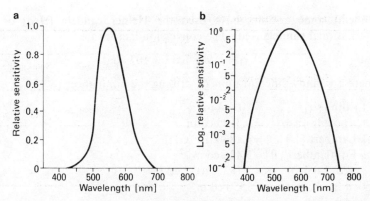

Fig. 1.2. Relative sensitivity of the bright-adapted human eye on a linear (a) and a logarithmic (b) scale (from Jahn, 1977).

radiation with a wavelength of 400 nm or 720 nm needs to be 1000 times as strong as at 555 nm in order to produce the same visual impression.

The SI unit of the luminous intensity I_v is 1 candela [cd]. This intensity is emitted by an area of 1.67×10^{-6} m^2 of a blackbody at the temperature of solidifying platinum (2045 K) perpendicular to its surface. The product of the luminous intensity and the solid angle is defined as the luminous flux (light flow) Φ with the dimension of lumen [lm]. The product of the luminous flux with time yields the (photometric) light quantity (luminous energy Q_v [lm s].

Analogously to the photon fluence, we can determine the luminous intensity per area unit defined as luminance L_v with the dimension [cd m^{-2}] = [nit]. When a light flow of 1 lm impinges on an area of 1 m^{-2} we obtain the illuminance E_v with the derived SI unit lux [lx]. Thus the visual unit of the illuminance is analogous to the energetic unit of the the fluence rate with the dimension [W m^{-2}]. The photometric fluence Φ_v is the product of illuminance and time [lx s].

In addition to the dimensions defined above, there is a realm of nonstandard units which should be avoided. Instead of a candela (new

TABLE 1.2   UNITS AND DIMENSIONS IN THE PHOTOMETRIC SYSTEM

	Symbol	SI unit
Luminous intensity	I_v	[cd]
Light flow	Φ_v	[lm] = [cd sr]
Light quantity	Q_v	[lm s] = [talbot]
Luminance	L_v	[cd m^{-2}] = [nit]
Illuminance	E_v	[lx] = [lm m^{-2}]
		= [cd sr m^{-2}]
Photometric fluence	Ψ_v	[lx s] = [lm s m^{-2}]

candela) some authors have used the Hefner candela (Hcd) or the international candela (Icd). The conversion factors are:

$$1 \text{ cd} = 0.981 \text{ Icd} = 1.107 \text{ Hcd}.$$

There are many different units describing the luminance:

1 Stilb [sb]	$= 10^4 \text{ cd m}^{-2}$
1 Apostilb [asb]	$= 0.3183 \text{ cd m}^{-2}$
1 Lambert [L]	$= 3.183 \times 10^3 \text{ cd m}^{-2}$
1 Footlambert [fl]	$= 3.426 \text{ cd m}^{-2}$
1 Skot [sk]	$= 3.183 \times 10^{-4} \text{ cd m}^{-2}$
1 cd in^{-2}	$= 1550 \text{ cd m}^{-2}$
1 cd ft^{-2}	$= 10.764 \text{ cd m}^{-2}$
1 cd cm^{-2}	$= 10^{-4} \text{ cd m}^{-2}$

The usage of the Hefner candela and the international candela results in additional dimensions such as $1 \text{ Hsb} = 0.903 \text{ sb} = 0.886 \text{ Isb}$.

Parallel units for the light flux are:

$$1 \text{ phot [ph]} = 10^4 \text{lx} = 1 \text{ lm cm}^{-2}$$
$$1 \text{ footcandle} = 10.764 \text{ lx}$$

Again, using different candelas increases the confusion so that the letters I or H have to be used for specificity: pH, Iph, Hph.

As indicated above, all visual units are based on the spectral sensitivity of the bright-adapted human eye $V(\lambda)$. We could also base them on the sensitivity of the dark-adapted eye and use the spectral sensitivity $V'(\lambda)$. Using this curve we define a dark luminance with the unit skot [sk] and a dark light flux with the unit nox [nx] (which corresponds to about 10^{-3} lx but which is of course based on the spectral sensitivity of the dark-adapted human eye).

Since the photometric system of units is defined by the specific spectral sensitivity of the human eye, we cannot simply convert between the photometric and the energetic systems unless we use monochromatic radiation. The conversion factor is the photometric radiation equivalent K_m of 673 lm W^{-1} which is unity at the optimum of the spectral sensitivity at 555 nm. At other wavelengths K_m needs to be multiplied with the sensitivity factor V, which can be found using Fig. 1.2. A simple example shows the conversion: a monochromatic irradiation at a wavelength of 620 nm and a fluence rate of 1 W m^{-2} corresponds to a light flux of

$$E_v = 1 \text{ W m}^{-2} \times 637 \text{ lm W}^{-1} \times 0.4 = 1683 \text{ lx}.$$

When instead of monochromatic radiation we use a wavelength band, we have to integrate over all wavelengths and have to take into consideration the specific sensitivity factors at each wavelength. Thus the conversion can produce quite different values depending on the wavelength composition of the radiation. A fluence rate of $1 \ W \ m^{-2}$ of 'white light' emitted from a tungsten filament lamp corresponds to about 29.7 lx. The major component of this radiation is infrared and therefore does not contribute to the measurement in the photometric system. The fluence rate of $1 \ W \ m^{-2}$ produced from a quartz-halogen bulb, which is used for example in modern low-voltage slide projectors, corresponds to about 302 lx and a fluorescence lamp produces about 826 lx.

1.5   THE LAWS OF OPTICS

This section is not intended to be substitute for a handbook on optics, but rather to facilitate the solution of practical optical problems using simple examples. Likewise, it is not meant to be an introduction to the use of optical instruments; instead we refer to the pertinent literature.

1.5.1   Geometrical optics

We start our considerations with an important characteristic of light which can be explained by assuming light to consist of either waves or particles. Every billiards player knows that the ball is reflected from the table edge at the same angle as the angle of incidence (provided there are no additional forces on the ball): angle of incidence = angle of reflection (Fig. 1.3). When we apply

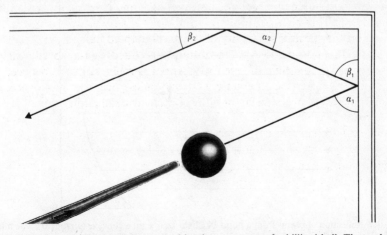

Fig. 1.3. The law of reflection demonstrated by the movement of a billiard ball. The angles of incidence α_1 and α_2 are identical with the angles of the reflected path β_1 and β_2, provided there are no additional forces on the ball (effet).

this law of reflection to our optical problems we need a 'point light source' and a concave mirror (Fig. 1.4). Of course, a technical light source is not an ideal point source and therefore we can only produce more or less parallel light.

The concave mirror represents a sector from the inner surface of a sphere with the radius Z–R. When we place a light source in the focus, F, which is halfway between Z and R, all rays are reflected which hit the mirror surface. The angle of incidence α between the light ray and the tangent to the surface of the sphere (which is perpendicular to the radius) is equal to the angle of reflection β between the tangent and the reflected beam. Of course we have to block the direct radiation from the light source to the object. All beams not hitting the mirror are lost. Parallel light is produced, however, only by rays close to the optical axis since for lateral rays the focal length is $< R/2$. Therefore the lateral rays deviate from the axis parallel to the axis. This phenomenon is called spherical aberration. This optical error is avoided by using a parabolic mirror, the surface of which is described by rotating a parabola around its axis. When we reverse the direction of the light rays we can concentrate parallel light in the focus.

The same effect can be obtained using a focussing lens. Before we discuss this problem we have to consider another property of light, refraction. A beam of light entering a transparent medium with a higher refractive index does not change its direction provided it hits the object perpendicular to its surface;

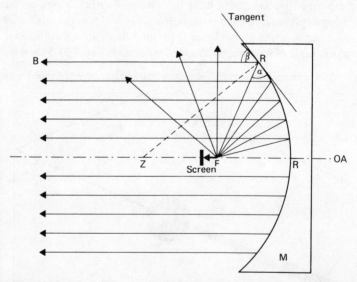

Fig. 1.4. Reflection of beams from a point-like light source on a concave mirror (M). The mirror surface is part of the inner surface of a sphere with a radius Z–R. The point source is located in the focal point F, halfway between Z and R on the optical axis OA. Direct beams from the light source to the image are blocked by a screen.

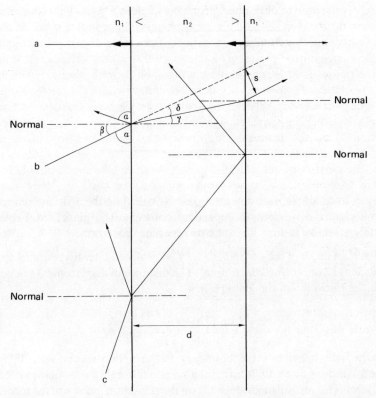

Fig. 1.5. Refraction of beams at the interface between two transparent substances such as air with a refractive index n_1 and glass with a refractive index n_2 and vice-versa. With perpendicular incidence, as in (a), there is no refraction but only partial reflection. When a beam hits the glass surface at an angle α to the surface as in (b), it is refracted by the angle δ towards the normal. This results in the new angle γ between the normal and the beam. During transition from glass (n_2) to air (n_1) the beam is refracted away from the normal. After passing through a glass plate with thickness d the beam is displaced parallel to itself by the distance s. At incidence angles larger than a critical angle (depending on the refractive index) we find total reflection as in (c).

only a certain fraction of the radiation is lost by reflection (Fig. 1.5a). When passing through a glass plate about 8% of the intensity of the incident beam is lost. When the rays hit the surface at an angle α with respect to the normal they are deflected towards the normal (Fig. 1.5b). The two angles are mathematically connected by Snell's law of refraction (1621)

$$\frac{\sin \beta}{\sin \gamma} = n_o.$$

The constant n_o is the relative refractive index defined as the ratio of the refractive indices between the two media. When we assume the transition of a

TABLE 1.3. REFRACTIVE INDICES OF SOME
MATERIALS AT 20°C AND $\lambda = 589$ nm

Hydrogen	1.0001
Air	1.0003
Water	1.33
Ethanol	1.37
Benzene	1.50
Crown glass	1.52–1.62
Flint glass	1.61–1.76
Carbon disulfide	1.63
Diamond	2.42

light ray from air (or vacuum) into glass we calculate the absolute refractive index n which is a (wavelength-dependent) constant of the glass used. Table 1.3 lists the refractive indices for some transparent substances.

Given an angle of incidence β, the ray is displaced laterally, parallel to the incident beam, toward the normal. The distance of displacement s depends on the thickness of the glass plate d

$$s = d \sin \beta \left(1 - \frac{\cos \beta}{\sqrt{n^2 - \sin^2 \beta}} \right)$$

When a light beam hits the boundary between two media with different refractive indices at an angle larger than a critical angle β it is totally reflected (Fig. 1.5c). This critical angle depends on the refraction index n of the medium. Total reflection occurs at a glass surface (with a refractive index of 1.6) when the angle of incidence is below 38.7°. The phenomenon of total reflection is used, for example, in a prismatic telescope.

In a medium with a gradual change in the refractive index a beam of light follows a curved path due to the continuously changing degree of refraction. This phenomenon can be observed in a salt solution with a gradually increasing concentration. It is also the basis for mirage images which can be observed at the boundary of air masses with different densities (due to temperature differences).

The reflected beam is partially polarized (see Section 3.4.3). The polarization is especially pronounced when the reflected and refracted beams are perpendicular to each other. For glass with a refractive index of 1.53 this is the case at an angle of incidence of about 57°. Both the incident beam and the reflected beam are in the plane of polarization (see Section 1.2).

After this excursion into the discussion of refraction we will consider its application in lenses. Figure 1.6 shows the path of a ray through a biconvex thin lens. Rays parallel to the optical axis are focussed in the focal point F. We can use such a lens to project the image of an object G at a distance a from the

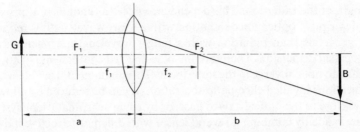

Fig. 1.6. Course of a beam through a biconvex, spherical, thin lens. Parallel beams from the object G are collected in the focal point F_2. At a distance b the enlarged, inversed, real image B can be projected onto a screen. The distances f_1 and f_2 are identical if the radii of the two lens surfaces are equal.

lens. The image B is real and inverted, and is produced at a distance b which can be calculated using the formula

$$\frac{1}{a} + \frac{1}{b} = \frac{1}{f}.$$

When the radii of the two spherical lens surfaces are identical $(r_1 = r_2)$ the focal length f equals the radius r. The image produced by the lens is a real image; it can be projected onto a screen in contrast to a virtual image, such as the image seen in a mirror. The reciprocal value of the focal length is the power of the lens measured in diopters. A lens with $f = 20$ cm has 5 diopters.

Also in lenses we are confronted with the problem of spherical aberration: rays close to the optical axis converge in a focal point further away from the lens than that for rays passing close to the outer edge. This aberration can be reduced by using an aperture which allows only central rays to pass.

Another common aberration is astigmatism, which occurs in spherical lenses when the object to be projected is far from the optical axis. The rays originating from the object hit the lens surface at an oblique angle, which prevents the production of a sharp, distinct image.

1.5.2 Physical optics

All phenomena described by geometrical optics can be explained without assuming light to be a wave. The radiation is considered to follow a linear path. Therefore, the theory proposed by Gauss and Abbe operates with rays drawn as geometric lines. The same principles can be applied to the optics of electrons and ions.

Coming back to the problem of refraction (see Fig. 1.5) we had defined the refractive index of a transparent material as a constant which depends on the wavelength of the radiation. This dependence poses a fundamental problem for the descriptive optics since radiation with a short wavelength is refracted by a larger angle than radiation with a longer wavelength, a phenomenon which is called dispersion. Therefore the blue and red components of a white light beam do not converge in the same focus. This fundamental problem of all lens projections is called chromatic aberration. It can be reduced by allowing only rays close to the optical axis to pass, or by using achromatic lens systems which are made by combining several lenses with different refractive indices.

The phenomenon of dispersion can also be used to separate the various wavelengths of white light in a prism (Fig. 1.7). When passing through a triangular prism with a prism angle α the beam is refracted by the angle γ:

$$\sin\left(\frac{\alpha+\gamma}{2}\right) = n \sin\left(\frac{\alpha}{2}\right).$$

The angle γ depends on the wavelength and therefore the white light beam is spread into its spectral components.

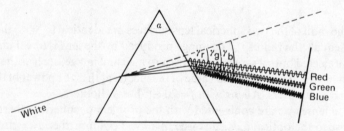

Fig. 1.7. Dispersion of white light by a prism with a prism angle α. Rays of different wavelengths (red, green and blue) are refracted by different angles (γ_r, γ_g, γ_b) (e.g. short-wavelength blue rays are refracted more than long-wavelength red rays).

The wave nature of light can also be demonstrated by the interferences produced by superimposing two waves. For this purpose we need two coherent light sources. Since atoms or molecules emit independent trains of waves, we cannot expect light from two light sources to be coherent. Therefore we have to derive the two light sources from a common one by reflection or refraction. Young (1803) used two narrow, adjacent slits as secondary light sources which were irradiated from a common source (Fig. 1.8). On a screen he observed an interference pattern of alternating dark and bright bands. With this experiment Young invalidated Newton's concept of light consisting of particles.

Seen from above, the two slits produce circular waves (Fig. 1.9). Two waves arising from these secondary light sources interfere constructively at a point on the screen when the phase difference between the two is an even multiple of $\lambda/2$. The first maximum at $Ox\lambda/2$ is called the zeroth fringe. At odd multiples of $\lambda/2$

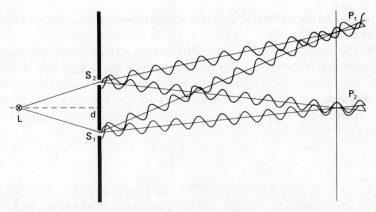

Fig. 1.8. Interference between beams from two coherent light sources (slits S_1 and S_2 at a distance d, illuminated by a common light source L). The waves arising from the two slits cancel each other out in P_1 and add to one another in P_2.

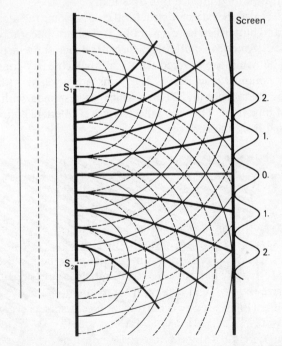

Fig. 1.9. Interference between two waves arising from the slits S_1 and S_2. When parallel wave fronts hit the slits, they act as secondary light sources from which semicircular (seen from above) waves originate. The wave peaks are shown as fine continuous lines and the troughs as broken lines. When the two waves meet on the screen with the same phases (such as two peaks or two troughs) the waves add up to maxima. When the waves have a phase difference of 180° (when, for example, a peak of one wave meets a trough in the other) they cancel each other. The maximum in the middle between the two slits is called maximum of zeroth fringe and all subsequent ones (in both directions) maxima of 1., 2., . . . n., order. The peaks are connected by heavy lines.

the two waves cancel each other, since a minimum in one curve coincides with a maximum of the other.

The phenomenon of interference is utilized to produce monochromatic light in a monochromator with a diffracting grating (see Section 3.5). A grating carries a large number of equidistant groves on a glass or metal surface. When monochromatic light with a wavelength of 500 nm falls on a grating with a grating constant of $d = 10^{-6}$ m (1000 lines per mm) we calculate an angle of 30° for the first fringe using

$$\sin \alpha = \frac{\lambda}{d}.$$

Since other wavelengths produce different deviation angles an incident white light beam is spread into its spectral components from which we can select a certain wavelength range. The halfband width of the monochromator depends on the slit width. The slit should not be too narrow, however, since that would decrease the transmitted fluence rate drastically.

Interference is also the basis for structural colors. Some red algae, as well as many beetles and butterflies, have a metallic, iridescent appearance. The wings of the Southern American butterfly *Morpho retenor* show a blue metallic iridescence. The wings are covered with scales which carry triangular veins supported by columns (Fig. 1.10). The veins are covered with horizontal parallel ridges spaced about 220 nm, which corresponds with half of the wavelength of blue light. Thus the reflected waves interfere constructively as light is reflected from the surfaces.

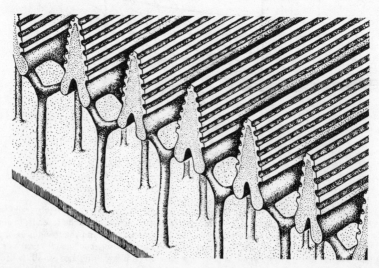

Fig. 1.10. Origin of the structural colors in the tropical butterfly *Morpho rhetenor* (male). The scales on the upper surface carry triangular veins supported by columns. The rods have parallel ridges at a distance of 220 nm so that blue light rays (440 nm) constructively interfere since the incident beam and the reflected beam have phase differences of $2n \times \lambda/2$ (from Nijhout).

1.6 BIBLIOGRAPHY

Textbooks and review articles

Anonymous. Units, Symbols Abbreviations. *Planta* **157**, 91–95 (1983)

Bell, C. J. and Rose, D. A. Light measurement and the terminology of flow. *Plant, Cell Environ.* **4**, 89–96 (1981)

Bergmann, L. and Schaefer, C. *Lehrbuch der Experimentalphysik. III. Optik.* Walter de Gruyter, Berlin (1966)

Coyle, J. D., Hill, R. R. and Roberts, D. L. (eds.) *Light, Chemical Change and Life: A Source Book in Photochemistry.* The Open University Press (1982)

Crawford, F. S. Jr. *Waves.* Berkeley Physics Course, Vol. 3. McGraw-Hill, New York and London (1965)

Flügge, J. *Praxis der geometrischen Optik.* Vandenhoeck und Rupprecht, Göttingen (1962)

Gerlach, D. *Das Lichtmikroskop.* Thieme, Stuttgart (1976)

Hartmann, K. M. Wirkungsspektrometrie. In: von Hoppe, W., Lohmann, W., Markl, H. and Ziegler, H. (eds), *Biophysik*, 2. Aufl. Springer, Berlin and Heidelberg, pp. 122–152 (1982)

Holmes, M. G. Radiation Measurement. In: Smith, H. and Holmes, M. G. (eds) *Techniques in Photomorphogenesis.* Academic Press, London, pp. 81–107 (1984)

Keitz, H. *Light Calculations and Measurement.* Macmillan, London (1971)

Loudon, R. *The Quantum Theory of Light*, 2nd edn, Oxford University Press (1983)

Michel, K. *Die Grundzüge der Theorie des Mikroskops.* Wiss Verlagsges., Stuttgart (1964)

Orear, J. *Grundlagen der modernen Physik.* Hanser, München (1973)

Pohl, R. W. *Optik und Atomphysik.* Springer, Berlin (1963)

Purcell, E. M. *Electricity and Magnetism.* Berkeley Physics Course, Vol. 2. McGraw-Hill, New York and London (1965)

Rabak, J. F. *Experimental Methods in Photochemistry and Photophysics*, Part I and II. J. Wiley & Sons, Chichester, New York, Brisbane, Toronto and Singapore (1982)

Rotter, F. Das internationale Einheitensystem in der Praxis. *Physik in unserer Zeit* **10**, 43–51 (1979)

Sears, F. W., Zemansky, M. W. and Young, H. D. *College Physics.* Addison-Wesley, London and Menlo Park, California (1975)

Smith, K. C. (ed.) *The Science of Photobiology.* Plenum Press, New York and London (1977)

Swanson, C. P. (ed.) *An introduction to Photobiology.* Prentice-Hall, London (1969)

US Department of Commerce, National Bureau of Standards: *The International System of Units.* Special Publication 330

Zimmer, H. G.: *Geometrische Optik.* Springer, Berlin (1967)

Further reading

Hartmann, K. M. and Cohnen-Unser, I. Analytical action spectroscopy with living systems: Photochemical aspects and attenuance. *Ber. Dtsch. Bot. Ges.* **85**, 481–551 (1972)

Jahn, J. Physikalische Grundlagen der Optoelektronik. *Funkschau* **49**, I 55, II, 71, III 163 (1977)

Nijnhout, H. F. The color patterns of butterflies and moths. *Sci. Amer. November*, 104–115 (1981)

Petzold, M. and Gensheimer, J. *Wie rechnet man SI-Einheiten um?* Hanser Verlag, Munich and Vienna (1974)

Rupert, C. S. Dosimetric concepts in photobiology. *Photochem. Photobiol.* **20**, 203–212 (1974)

Rupert, C. S. Uniform terminology for radiations. *Photochem. Photobiol*, **28**, 1 (1978)

Rupert, C. S. and Latarjet, R. Toward a nomenclature and dosimetric scheme applicable to all radiations. *Photochem. Photobiol* **28**, 3–5 (1978)

Sacklowski, A. *Einheitenlexikon.* Dtsch. Verlags-Anstalt, Stuttgart (1973)

Tape, W. The topology of mirages. *Sci. Amer.* **252**, 100–106 (1985)

2   Light sources

The sun emits the whole electromagnetic spectrum; however, only part of it reaches the surface of the earth. Visible light as well as the adjacent ultraviolet and infrared bands, which are invisible to the human eye, can also be produced with artificial light sources. The photobiologist is specially interested in the visible part of the electromagnetic spectrum, since photoreceptors of microorganisms (see Chapters 8 and 12), plants (see Chapters 9 and 11) and animals (see Chapter 13) absorb light in this range. Since the various photoreceptors differ in their absorption ranges it is important to know the spectral distribution of natural and artificial light sources.

In addition to visible light, the invisible ultraviolet radiation is biologically extremely effective, since a number of biomolecules, specially nucleic acids and proteins, absorb in this range. The high energetic UV radiation causes photochemical changes in the structures which carry the genetic information, thus causing permanent damage or the death of the organism (see Chapter 14).

Visible light and the adjacent wavelength bands can be produced by several processes:

1. *Thermal radiation.*   The sun and many artificial light sources such as incandescent lamps or photo flashbulbs emit radiation due to a thermal excitation.
2. *Electric discharge in metal vapors and gases.*   In this group we find sodium and mercury vapor lamps as well as hydrogen and xenon lamps. The primary process in a fluorescent lamp is also an electric discharge which induces a secondary excitation of the fluorescent substances which coat the inside of the tube. In recent years, lasers have been used in photobiological and photochemical research. Their action is also based on energetic excitation and discharge in rare gases, crystals or dyes.

2.1   NATURAL RADIATION

When we heat a blackbody to about 500°C it emits infrared radiation (heat). When we increase the temperature beyond 500°C it emits light in addition to heat. According to Planck's law of radiation and Wien's law of displacement, the spectral energy distribution of a blackbody is defined by its temperature; the higher the temperature the higher its emission maximum and the shorter the emitted wavelengths.

The term 'blackbody' is misleading: it looks black at room temperature because it absorbs most of the radiation, but at a temperature of 1000°C it glows in white heat. According to Kirchhoff's law a blackbody shows the highest possible emission.

The sun is a natural light source which produces thermal radiation. It obtains its energy by nuclear fusion in which matter is transformed into energy which is emitted into space as electromagnetic radiation. The spectral energy distribution is equivalent to that of a blackbody at about 5700°C (Fig. 2.1).

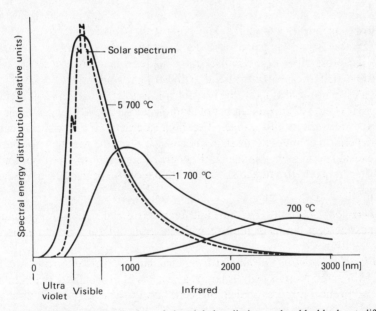

Fig. 2.1. Spectral energy distribution of the global radiation and a blackbody at different temperatures (schematic) (from Weischet, 1979, modified and extended).

The total extraterrestrial radiation outside the atmosphere has a fluence rate of about 1350 W m^{-2} averaged over the year and is called solar constant. However, due to the changing distance between the sun and the earth the fluence rate varies with the season. In January one measures a fluence rate of about 1410 W m^{-2} and in July of about 1310 W m^{-2}. The total visible radiation amounts to about 50% of the solar constant, the UV radiation to about 8% and the remainder is heat (Table 2.1).

TABLE 2.1　Spectral distribution of the extraterrestrial solar radiation and the global radiation measured as [W m^{-2}] for a solar azimuth of 90° (after CIE publication No. 22 and data from Schulze, in Kiefer, 1977).

Wavelength [μm]	Extraterrestrial solar radiation [W m^{-2}]		Global radiation [W m^{-2}]		Percentage absorption
<0.28	8		0		100
0.28–0.32	22 ⎱		5 ⎱		
0.32–0.36	42 ⎰ 110		27 ⎰ 68		38.2
0.36–0.40	46		36		
0.40–0.44	67		56		
0.44–0.48	80		73		
0.48–0.52	77		71		
0.52–0.56	71		65		
0.56–0.60	68		60		
0.60–0.64	64	638	61	580	9.1
0.64–0.68	59		55		
0.68–0.72	55		52		
0.72–0.76	50		46		
0.76–0.80	47		41		
0.80–1.0	182		156		
1.0–1.2	120	382	108	329	13.9
1.2–1.4	80		65		
1.4–1.6	58		44		
1.6–1.8	40		29		
1.8–2.0	26	183	20	143	21.9
2.0–2.5	38		35		
2.5–3.0	21		15		
>3.0	29		—		100

2.1.1　Radiation climate

The spectral distribution of the radiation on the surface of the earth is called global radiation. Considerable fractions of the extraterrestrial solar radiation are absorbed by the atmosphere (see Table 2.1). The short wavelength range below 290 nm and the long wavelength range above 3000 nm are quantitatively absorbed. Assuming a cloudless sky and a solar azimuth of 90°, only 23% of the UV-B, 72% of the UV-A and 91% of the visible extraterrestrial radiation reach the surface of the earth. At lower azimuths or with clouded skies, considerably more radiation energy is absorbed (Fig. 2.2).

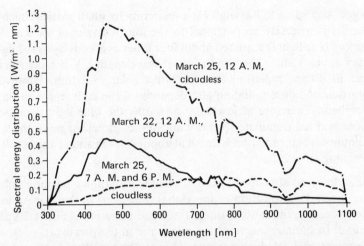

Fig. 2.2. Spectral energy distribution of the global radiation on 22 March 1981 (cloudy) and 25 March 1981 (clear sky) at different times of the day in the wavelength range 300–1100 nm at Karlsruhe, FRG (49°N).

2.1.1.1 GLOBAL RADIATION AND ITS COMPONENTS

Global radiation is defined as the sum of the direct solar radiation and the diffuse scattered radiation from the sky which hits an area per time unit. In addition to selective absorption of specific wavelength bands–especially by ozone, carbon dioxide and water vapor in the atmosphere–global radiation is altered by diffuse scattering. The latter is caused predominantly by gas molecules in the atmosphere, water droplets and ice crystals in clouds (10^{-6} to 10^{-4} m in diameter) and by aerosols ($< 10^{-7}$ m in diameter) which are composed mainly of dust, smoke particles and vapors. Depending on the solar azimuth, selective absorption and reflection can reduce solar radiation by up to 70% (Table 2.2).

Clouds have drastic effects on global radiation. Cloud cover can reduce the fluence rate by up to 80% (Fig. 2.2). Under these conditions the amount of diffusely scattered radiation is especially high and the spectral distribution

TABLE 2.2   GLOBAL RADIATION AS A PERCENTAGE OF THE SOLAR CONSTANT I_o ($= 1350$ W m^{-2}) UNDER CLOUDLESS SKY IN RELATION TO THE AZIMUTH OF THE SUN (FROM SCHULZE, IN KIEFER, 1977)

Solar azimuth	5°	10°	30°	60°	90°
High mountains	55	65	80	86	87
Plains	45	55	73	82	83
Large cities	37	47	67	77	79
Industrial areas	30	39	60	71	74

changes. According to Rayleigh's law scattering by small particles such as gas molecules is inversely proportional to the fourth power of the wavelength. Therefore blue light is scattered about four times as effectively as red light and the sky appears blue. The band of short-wavelength UV-B radiation is also higher in diffuse radiation than in direct solar radiation. However, the proportion of diffuse radiation also depends on the path length through the atmosphere. Therefore at low solar azimuths the blue light component is reduced and red dominates (sunrise and sunset glow). At noon the spectral maximum of solar radiation is found at about 480 nm and at about 680 nm in the evening (Fig. 2.2).

In addition to atmospheric conditions the geographic position influences the spectral composition of the global radiation. In the tropics, under a cloudless sky, the ratio of direct to diffuse radiation is about 5:1 all year round. In summer one finds a similar ratio at temperate latitudes while it decreases to 1.5:1 during winter. Due to the mostly cloudless skies, the mean global radiation is highest in deserts (8.4×10^9 m^{-2} per year while central Europe receives about 4.2×10^9 J m^{-2} per year).

The global radiation reaching the surface of the earth is either absorbed or reflected. The degree of reflection depends on the properties of the surface. White surfaces such as snow or ice layers have a pronounced reflection. The ratio between reflected and incident radiation energy is defined as albedo and usually expressed as a percentage. Smooth surfaces such as water or concrete show an albedo between 7% and 27% (albedo quotient 0.07 to 0.27). Areas covered with vegetation have a mean albedo of about 16%. The unreflected fraction of the radiation is absorbed by the soil or by plants. Green plants convert the light energy into chemical energy, heat or fluorescence (see Chapter 5). Absorption of radiation by the soil generates heat. However, this heat is radiated back into the atmosphere where it is mostly reabsorbed by carbon dioxide or water vapor. Thus, this energy is trapped in the atmosphere close to the surface. By this mechanism, the CO_2 concentration influences the temperature of the earth. In addition to the spectral composition of the global radiation, this so-called greenhouse effect greatly influences the climate and thus the development of the vegetation.

2.1.1.2   UV RADIATION AND THE STRATOSPHERIC OZONE LAYER

Carbon dioxide and water vapor mainly absorb infrared radiation in the atmosphere. Carbon dioxide selectively absorbs in various bands between 2.3 μm and 3 μm, between 4.2 μm and 4.4 μm and between 12 μm and 16 μm. Water vapor absorbs between 2.5 μm and 8 μm and absorbs all infrared radiation with wavelengths above 14 μm. The strong absorption of the damaging UV radiation with wavelengths below 320 nm is mainly due to the ozone layer (O_3) in the stratosphere at an altitude between 20 and 50 km. In the following we will describe the relationship of the ozone layer and the UV radiation in more detail since the filter function of the ozone layer in decreasing the level of damaging UV radiation is of enormous importance for the survival of life on this planet.

Ozone is produced after the photolytic cleavage of oxygen by short wavelength radiation. In a slower process, UV radiation with longer wavelengths again splits the ozone which eventually produces molecular oxygen:

$$O_2 + h\nu(<200 \text{ nm}) \rightarrow O + O$$
$$O + O_2 + N \rightarrow O_3 + M$$
$$O_3 + h\nu(<300 \text{ nm}) \rightarrow O_2 + O$$
$$O + O_3 \rightarrow 2\,O_2$$

N = nitrogen as activating partner

Due to UV absorption by stratospheric ozone, UV radiation with wavelengths below 285 nm is virtually completely prevented from reaching the surface of the earth. The UV level measured today will drastically increase over the next few decades if the stratospheric ozone layer is reduced in size by halogenated hydrocarbons which are presently used as propellents in spray cans and as foaming agents or coolants in refrigeration. The most important chlorinated fluorhydrocarbons are $CFCl_3$ (F11), CF_2Cl_2 (F12), known under the brand names of Frigen and Freon, as well as the chlorinated hydrocarbons CH_3Cl, CCl_4 and CH_3CCl_3.

In addition to these substances, nitrogen oxides can affect the ozone layer. These are produced by combustion processes, by volcanic or microbial activity. Chlorine or nitrogen oxide radicals formed photochemically from the gases mentioned above substitute atomic oxygen and react catalytically with ozone producing molecular oxygen. The best-known chlorine and nitrogen oxide radical reactions are:

$$Cl + O_3 \rightarrow ClO + O_2$$
$$ClO + O \rightarrow Cl + O_2$$
$$NO + O_3 \rightarrow NO_2 + O_2$$
$$NO_2 + O \rightarrow NO + O_2$$

$$\overline{O + O_3 \rightarrow 2\,O_2}$$

Other factors affecting the ozone layer are mixing, transport and diffusion of gases, rhythmic changes in the solar spot activity and stratospheric temperature changes. The ozone layer is also strongly dependent on geographical latitude. At the same time of the year, there is an ozone concentration of 245 D (1 Dobson = 0.001 cm bar, not an SI unit) at the equator and one of 350 D in the Northern Hemisphere. Due to the lower ozone concentration the UV irradiation is higher at the equator than at temperate latitudes.

The factors discussed above hinder an exact prediction of the ozone destruction by gaseous pollutants. But provided that the ozone layer is reduced we will face an increase in the UV-B fluence rate and a shifting of the solar cutoff to shorter wavelengths. Both effects increase the biological effectivity of the radiation (Chapter 14) on all exposed organisms–microorganisms, plants, animals and men.

2.2   ARTIFICIAL RADIATION

Artificial radiation is used for lighting, but also to rear plants in greenhouses or growth chambers. Plants thrive best when the artificial light sources emit enough energy in the blue and red regions of the spectrum which correspond with the absorption maxima of the photosynthetic pigments. Due to differences in the spectral distribution, different artificial light sources may have different effects on organisms. Both thermal and electric discharge radiation sources are being used.

2.2.1   Thermal radiation sources

The incandescent lamp is still the most commonly used light source at home. An electric current is passed through a tungsten filament placed in a vacuum or in an inert gas such as nitrogen, argon or krypton and causes it to glow. As stated above, the energy distribution of a blackbody depends on its temperature (Fig. 2. 1). At a temperature of 2700°C which has been measured in a tungsten filament (melting point 3380°C), a considerable amount of infrared is emitted in addition to visible light. Therefore, in photobiological experiments the infrared components are usually removed by inserting into the light beam heat absorbing filters (KG filters, Schott & Gen., Mainz) or water cuvettes. The heat problem can also be solved by using dichroic mirrors which reflect the visible light but allow the infrared radiation to pass (cold light mirrors).

Since 1959, incandescent lamps have been produced in which a tungsten halogen cycle takes place (e.g. Osram, halogen star). Vaporized tungsten does not condense on the inside of the bulb but reacts with iodine or bromine forming tungsten iodide or bromide which returns to the tungsten filament by convection and is split again into halogen and tungsten which is then deposited on the filament. These lamps produce higher fluence rates than normal light bulbs, since tungsten is not deposited on the inside of the glass bulb and occludes the light. Halogen lamps are utilized in increasing numbers in industrial applications.

Unlike fluorescence lamps, incandescent lamps have the major advantage that they do not need a starter and can be installed quickly. In addition, their fluences rates are rather constant over time. Some disadvantages for plant growth are the low yield of visible light compared to the electric energy used, the high fraction of infrared radiation and the high sensitivity to fast changes in temperature. In addition, changes in the voltage result in alterations of the spectral distribution and the lifetime of incandescent lamps is relatively short. Some of these disadvantages are not found in gas discharge lamps which operate by a totally different principle.

2.2.2   Luminous discharge lamps

In a discharge lamp the electric current flows between two electrodes through metal vapors or an ionized gas. The electrons emitted from the cathode hit the metal or gas atoms and excite an electron to a higher energy orbital. After excitation the electrons spontaneously return to their ground state and emit the energy difference in the form of fluorescence (see Chapter 5). The spectral energy distribution is determined by the nature of the metal vapor or gas as well as by the pressure inside the lamp.

2.2.2.1 LOW-PRESSURE DISCHARGE LAMPS

As early as 1929 the first low-pressure discharge lamp using vaporized sodium had been developed and used for street illumination. At a pressure of only a few Torr (1 Torr = 133.322 Pa) sodium emits at wavelengths of 589.0 nm and 589.6 nm, which is yellow light. Electric discharge in mercury at low pressure produces resonance lines at 184.9 nm and 253.7 nm. The shorter of the two wavelengths is absorbed by the quartz tube. Low-pressure mercury lamps are used for sterilization.

In principle, fluorescence lamps are also luminous discharge lamps. They use the UV radiation of the mercury discharge to induce the fluorescence of a phosphor (Fig. 2.3). After excitation the phosphor radiates in the yellow and red spectral regions. By choosing specific phosphors, the emission bands can

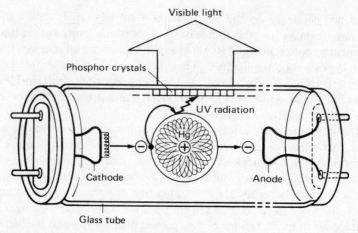

Fig. 2.3. Schematic diagram of a fluorescent lamp during operation (for details see text).

also be shifted into the UV-A or UV-B bands (e.g. Philips TL 40/12). In order to cut out the harmful mercury excitation radiation, normal fluorescent tubes are made of UV-absorbing glass. From the multitude of types on the market we have chosen a few especially useful ones for growing plants such as Fluora and Daylight. Their spectral energy distributions are shown in Fig. 2.4. In contrast to the line spectrum of the mercury lamp, the hydrogen and deuterium lamps used in spectrophotometry emit in a broad continuous band in the UV range (Fig. 2.4).

2.2.2.2   HIGH-PRESSURE DISCHARGE LAMPS

High-pressure discharge lamps are defined by the high pressure of about 10^6 Pa($= 1$ MPa) inside the lamp under which the electric discharge takes place in metal vapors or gases. In sodium vapor lamps at high pressure the resonance lines at about 590 nm are broadened. In mercury vapor lamps the emission at the wavelength of 254 nm is strongly decreased in favor of the spectral lines at 366 nm, 546 nm and 588 nm. These light sources are used as high intensive UV-A radiation sources in photochemotherapy of skin diseases (Psoriasis vulgaris, acne vulgaris, etc.) and for tanning. Addition of metal halides such as tin iodide produces a multitude of spectral lines or a broad continuous band in the visible range. Metal halogen lamps are used in TV studios and sport stadiums. Their yield is several times higher than that of incandescent lamps.

At high temperature, rare gases also emit light after electric discharge. Xenon high-pressure discharge lamps have a continuous spectrum which

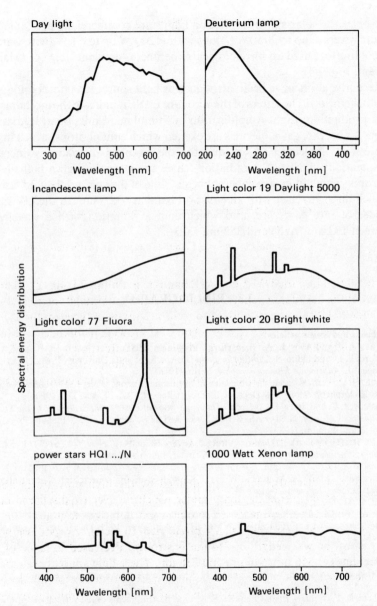

Fig. 2.4. Spectral energy distribution of light from incandescent lamps, fluorescence and electric discharge lamps (from Osram) as well as daylight (see Fig. 2.2).

resembles the solar spectrum. Xenon lamps are commercially available with input powers of up to 20 kW. Lamps with < 200 W up to 6 kW (with water or air cooling) are used for physiological experiments on plants (e.g. XBO lamps, Osram; Fig. 2.4).

Recently, lasers have been introduced as light sources in photobiology and photochemistry. The letters of the acronym LASER indicate the mechanism of light production (Light Amplification by Stimulated Emission of Radiation). In gas lasers, rare gas molecules are excited which emit photons, which in turn excite new molecules. Consequently the number of photons increases exponentially. Part of the radiation leaves the laser through a half-silvered mirror as a high intensive coherent beam. One of the commonly used lasers is the helium–neon laser with an emission band at 632.8 nm. In the UV range, halides of rare gases are used which emit a radiation with a wavelength between 193 nm (ArF) and 352 nm (XeF).

2.3   BIBLIOGRAPHY

Textbooks and review articles
Barbato, J. P. and Ayer, E. A. *Atmospheres*. Pergamon (1981)
Bickford, E. D. and Dunn, S. *Lighting for Plant Growth*. Kent State University Press, USA (1972)
Blüthgen, J. *Allgemeine Klimageographie*. Berlin (1966).
Dütsch, H. U. Vertical ozone distribution and tropospheric ozone. *Proc. NATO Adv. Study Inst. on Atmospheric Ozone*. US Dept. of Transport Report FAA-EE-80-20, 7 (1980)
Earnshaw, J. C. and Steer, M. W. (eds). *The Application of Laser Light Scattering to the Study of Biological Motion*. NATO ASI Series, Series A: Life Sciences, Vol. 59. Plenum Press, New York and London (1983)
Fabian, P. *Atmosphäre und Umwelt*. Springer, Berlin, Heidelberg, New York and Tokyo (1984)
Kiefer, J. *Ultraviolette Strahlen*. Walter de Gruyter, Berlin and New York (1977)
Lubinska, A. Ozone depletion. Europe takes a cheerful view. *Nature* **313**, 727 (1985)
Prather, M. J., McElroy, M. B. and Wofsy, S. C. Reductions in ozone at high concentrations of stratospheric halogens. *Nature* **312**, 227–31 (1984).
Smith, K. C. *The Science of Photobiology*, 2nd edn. Plenum, New York (1986)
Weischet, W. *Einführung in die Allgemeine Klimatologie*. G. B. Teubner, Stuttgart (1979)
World Meteorological Organization (WMO): *The Stratosphere 1981, Theory and Measurements*. WMO Global Ozone Research and Monitoring Project, Report No. 11, Geneva (1985)

Further reading
Blumthaler, M., Ambach, W. and Canaval, H. Seasonal variation of solar UV-radiation at a high mountain station. *Photochem. Photobiol.* **42**, 147–152 (1985)
Borchers, R. P., Fabian, P. and Penkett, S. A. First measurements of the vertical distribution of CCl_4 and CH_3CCl_3 in the stratosphere. *Naturwiss.* **10**, 14–16 (1983)
Chameides, W. L. and Davis, D. D. Chemistry in the troposphere. *Chem. Eng. News* **60–40**, 39 (1982)
Chapman, S. A theory of upper atmospheric ozone. *Mem. Roy. Meteorol. Soc.* **3**, 103 (1930)
Druckschrift der Osram GmbH: *Licht für Innen und Außen*. Munich April (1981)
Fabian, P., Borchers, R., Flentje, G., Matthews, W. A., Seiler, W., Giehl, H., Bunse, K., Müller, F., Schmidt, U., Volz, A., Khedim, A. and Johnen, F. J. The vertical distribution of stable gases at midlatitudes. *J. Geophys. Res.* **86**, 5179 (1981)

Fabian, P., Borchers, R., Penkett, S. A. and Prosser, N. J. D. Halocarbons in the stratosphere. *Nature* **294**, 733 (1981)

Farman, J. C., Gardiner, B. G. and Shanklin, J. D. Large losses of total ozone in Antarctica reveal seasonal ClO_x/NO_x interaction. *Nature* **315**, 207–210 (1985)

Froidevaux, L. and Yung, Y. L. Radiation and chemistry in the stratosphere: sensitivity to O_2 absorption cross section in the Herzberg continuum. *Geophys. Res. Lett.* **9**, 854 (1982)

Gille, J. C., Smythe, C. M. and Heath, D. F. Observed ozone response to variations in solar ultraviolet radiation. *Science* **225**, 315–317 (1984)

Green, A. E. S., Cross, K. R. and Smith, L. A. Improved analytical characterization of ultraviolet skylight. *Photochem. Photobiol.* **31**, 59–65 (1980)

Maugh II, T. H. What is the risk from chlorofluorocarbons? *Science* **223**, 1051–1052 (1984)

Rosenthal, F. S., Safran, M. and Taylor, H. R. The ocular dose of ultraviolet radiation from sunlight exposure. *Photochem. Photobiol.* **42**, 163–171 (1985)

Schulze, R. UV-Strahlenklima. In: Kiefer, J. (ed.) *Ultraviolette Strahlen*. Walter de Gruyter, Berlin and New York (1977)

Weiss, R. F., Bullister, J. L., Gammon, R. H. and Warner, M. J. Atmospheric chlorofluoro-methanes in the deep equatorial atlantic. *Nature* **314**, 608–610 (1985)

3 Absorption and wavelength selection

Chapter 1 discussed the calculations by which we can determine the energy of radiation which strikes an area per time unit (fluence rate). It is obvious that only that part of the radiation which is absorbed can be photochemically effective (Grotthuss and Draper law, 1872).

3.1 ABSORPTION OF RADIATION

Matter is not homogeneously distributed, but is concentrated in subatomic particles. Therefore not all photons hit an electron but a certain fraction of the incident energy I_0 passes through the object. This fraction I_t is called transmitted radiation, and the transmittance T is a measure for the transparency of an object.

$$T = \frac{I_t}{I_0}$$

We can also use the absorbance A as an inverse measure of transmittance.

$$A = \log \frac{I_0}{I_t}$$

In the older literature we find the symbol E (extinction) for absorbance. Both transmission and absorbance of light by a substance depend on the wavelength of the radiation. Therefore the absorption properties of a substance are described by its absorption spectrum. The absorbance of a substance or solution also depends on its optical pathlength l.

$$A_\lambda = m_\lambda l$$

where m_λ is the absorption module. The absorption of a dissolved substance also depends on its concentration c.

$$m_\lambda = \varepsilon_\lambda c$$

32

The wavelength-dependent constant ε_λ is defined as molar absorption coefficient (extinction coefficient). Combining the two equations yields the Lambert Beer law.

$$A_\lambda = \varepsilon_\lambda c l$$

Knowing the thickness of a cuvette l we can calculate the concentration of a dissolved substance from its absorbance at a given wavelength. Usually we choose a wavelength in the absorption maximum of the substance. The molar absorption coefficient ε_λ of hemoglobin at a wavelength of 430 nm is about $126 \, mg^{-1} \, ml^{-1}$. When we use a 1 cm cuvette and measure an absorbance of 0.4 we calculate a concentration of the hemoglobin solution of

$$c = \frac{A_\lambda}{\varepsilon_\lambda l} = \frac{0.4}{126} = 3.17 \, \mu g \, ml^{-1}.$$

Impurities of the substances, associations between molecules and interactions between the dissolved substances and the solvent can cause deviations from the Lambert Beer law.

Up to now we have neglected the reflected fraction I_r. At each transition between media with different refractive indices a fraction of the beam is reflected (see Section 1.5.1). Thus the incident radiation is split into a transmitted, an absorbed and a reflected fraction.

$$I_0 = I_t + I_a + I_r$$

When we consider a turbid suspension of particles instead of a clear solution, part of the incident light is diffusely scattered in all directions. The fraction of the scattered radiation is not a constant but depends on the wavelength (Rayleigh's law, 1899). The fraction of the scattered radiation is far higher for short wavelengths than for longer ones (see Chapter 2). This factor needs to be taken into account for the measurement of absorption spectra of suspensions.

3.2 QUANTUM YIELD

Not all absorbed quanta induce a photochemical reaction. Therefore we define a quantum yield Φ which is in the range between zero and one.

$$\text{Quantum yield } \Phi = \frac{\text{number of product molecules}}{\text{number of absorbed quanta}}$$

Using the physical considerations described in the previous section we want to determine for our practical example what fraction of the incident light quanta induces a photochemical response. We want to find out what fraction of the radiation from a projector is used for photosynthesis of a leaf. At first glance this task looks easier than it is. First, we have to know the number of incident quanta. Using a thermopile (see Section 4.2) we can measure the fluence rate of the radiation that strikes the leaf surface. Knowing the area we can determine the incident flux. We know the energy of a quantum at a given wavelenth (see Section 1.3) but we cannot expect the light source to emit the same number of quanta or the same energy at each wavelength. The wavelength-dependent energy distribution of a light source is characterized by its emission spectrum (see Section 2.2; Fig. 3.1). Such a spectrum can be determined by dividing the relevant wavelength range into small bands and measuring the fluence rate in each individual band. The integration over all these bands yields the total fluence rate. This measurement can be performed automatically with a spectroradiometer which scans the fluence rate at each wavelength. Because we know the energy of a quantum at each wavelength we can calculate the number of incident quanta from the emission spectrum of the light source.

In addition to the visible and photosynthetically effective radiation, the light source emits some infrared radiation. In order to decrease the heat stress on the object, we eliminate the infrared radiation with a heat-absorbing filter. The product of the light source emission spectrum and the transmission of the heat-absorbing filter yields the fluence rate of the radiation impinging on the leaf (Fig. 3.1).

Using a spectrophotometer (see Section 4.3), we can measure at each wavelength the fraction of the radiation transmitted through a leaf. The transmitted radiation can be subtracted from the incident radiation. The

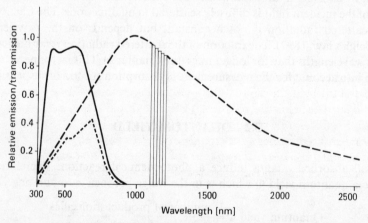

Fig. 3.1 Emission spectrum of an incandescent lamp with a color temperature of 2800 K (long dashed line, from Bergtold) and transmission spectrum of an infrared absorbing filter (KG3, Schott & Gen., continuous line). The product of the two spectra (short dashed line) represents the wavelength distribution of the transmitted radiation. In the wavelength range between 1100 and 1200 nm the emission is divided into small bands, the fluence rate of which is measured separately (for details see text). Abscissa: wavelength in nm, ordinate: emission and transmission drawn on a linear scale normalized to 1.

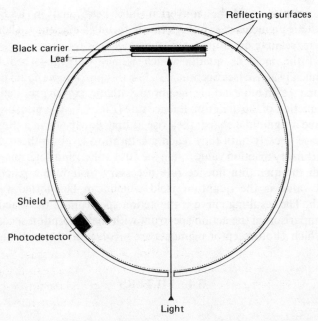

Fig. 3.2. Section through an Ulbricht sphere to measure the reflected and the scattered radiation. The measuring beam hits the sample (e.g. a leaf) through a window. Most of the transmitted radiation is absorbed by the black carrier. The reflected and scattered radiations undergo multiple reflection at the strongly reflecting surfaces (dotted, coated for example with Eastman white reflectance paint, No. 1415) until they are absorbed by the photodetector. The direct beam from the leaf to the photoreceptor is blocked by a shield (from Heath, modfied).

fractions of the reflected and the scattered radiation can be measured with an Ulbricht sphere (Fig. 3.2). The leaf is mounted inside a sphere, covered with white lusterless paint on the inside, and it irradiated through a small hole. Most of the light transmitted through the leaf is absorbed by a black surface behind it. The light reflected from the leaf is reflected repeatedly by the sphere until it strikes the photodetector. By scanning the wavelength we can determine a reflectance spectrum. One of the possible errors of this method is that a light quantum reflected from the leaf the first time can be absorbed by the leaf rather than by the photodetector during multiple reflection inside the sphere.

3.3   ACTION SPECTROSCOPY

The methods described above allow us to determine the approximate number of quanta absorbed by the object at each wavelength. The next problem is what fraction of the absorbed quanta is photosynthetically active. Some of the absorbed quanta could be converted into heat, and in addition the photosynthetic pigments may not operate with 100% efficiency. Therefore we use the term 'relative quantum yield', which defines for each wavelength the fraction of the incident quanta which is not only absorbed but also photochemically active (see Section 3.2). For this purpose we could determine the reaction rate (in our example the photosynthetic oxygen production of the leaf) as a function of the quantum fluence rate (Fig. 3.3). The measured curves usually have a sigmoidal shape: they rise at first slowly above a threshold T, then increase linearly until they reach a saturation level. In these curves we select an arbitrary reaction value (often the 50% value) and determine for each wavelength the quantum fluence rate necessary to induce a reaction. The reciprocal value is the quantum yield which can be plotted versus the wavelength. The resulting curve is the action spectrum of the reaction (Fig. 3.4). A comparison of the action spectrum with the absorption spectrum can indicate which photoreceptor pigments are involved in the reaction.

3.4   FILTERS

The emitted radiation of a light source often does not meet the experimental requirements, and it is usually modified by inserting filters. We can either

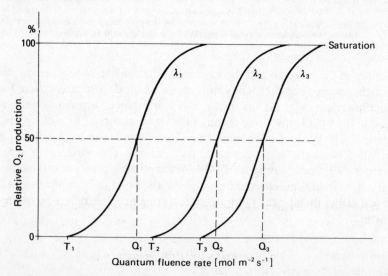

Fig. 3.3. Theoretical dependence of the photosynthetic oxygen production rate (in mol s^{-1}, as a percentage of the saturation value = 100%) on the quantum flux density (abscissa in mol s^{-1} m^{-2}) at the wavelengths λ_1, λ_2 and λ_3. The curves start at the threshold values T_1, T_2 and T_3 and reach a half-maximal reaction rate at the quantum flux densities Q_1, Q_2 and Q_3, which are used to calculate an action spectrum (Fig. 3.4).

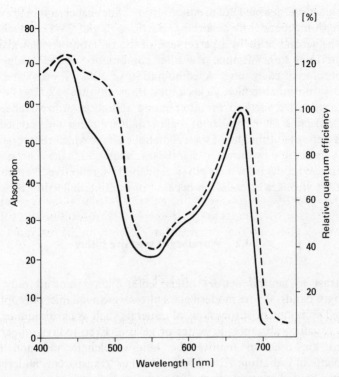

Fig. 3.4. Relative quantum efficiency of photosynthetic oxygen production (continuous line, right ordinate as percentage of the value measured at 670 nm) in comparison with the absorption spectrum (left ordinate) of a leaf (broken line) (after Ray (1963) from Mohr, modified).

decrease the fluence rate uniformly over a specified wavelength range (neutral density filters) or we can selectively suppress specific wavelengths (color filters).

3.4.1 Neutral density filters

Neutral density filters decrease the fluence rate of the incident radiation either by absorption or by reflection. A completely wavelength-independent attenuation can only be achieved by using nets or meshes which transmit a certain percentage of the incident radiation, which is proportional to the percentage of the open areas. One disadvantage of this method is that the fluence rate may not be uniform over the object. Furthermore nets or meshes cannot be stacked to obtain a higher absorption since the interference between the hole patterns causes the formation of so-called Moirè patterns; the degree of absorption is not predictable. Both disadvantages are avoided using neutral density glass filters which are produced from a fine suspension of gray (that is wavelength independently absorbing) particles in glass. When stacking two filters, the second absorbs a percentage of the radiation transmitted by the first. The total transmittance of a filter combination is the product of the transmittance of each filter. A combination of 10% ($T=0.1$) and a 25% ($T=0.25$) transmission filter yields a total transmittance of 2.5% ($T=0.025$).

The alternative method for attenuating radiation utilizes reflection. A neutral reflective filter consists of a glass or fused silica slide coated with a vacuum-deposited thin film of several metallic alloys which reflect part of the incident radiation without a significant change in the chromicity. A disadvantage results from their physical properties: reflective filters cannot be stacked for stronger attenuation because of multiple reflection between the filters.

3.4.2   Wavelength-selective filters

In contrast to neutral density filters, color filters transmit only certain wavelength bands. Before modern glass filters were available, photophysiologists used solutions or suspensions of minerals such as chromium or copper salts or organic substances in water or gelatin. Even today, copper sulfate solutions are used to remove the long-wavelength red and infrared components of radiation. Also glass and other 'transparent' materials only transmit a limited spectral range. If the material is intended to transmit UV radiation, lenses and prisms made from quartz are used.

Wavelength-selective absorbing glass filters can be divided into two groups: band filters and cutoff filters. Band filters are usually made of solutions of simple or complex ions such as nickel, cobalt or chromium oxide in glass. These substances absorb and transmit certain broad-wavelength bands. Some examples of band filters are shown in Fig. 3.5. In cutoff filters the transmission characteristics depends on the (submicroscopic) size of particles such as sulfur, gold or cadmium salts dispersed in glass. The term 'cutoff filters' describes the optical characteristics: the short-wavelength radiation of the incident beam is blocked below a relatively sharp edge (Fig. 3.6). The filters are usually named according to the wavelength at which they transmit 50% of the incident radiation. (An OG 515 is an orange glass which transmits 50% at 515 nm, Schott & Gen., Mainz, FRG.) The transmission curves indicate that radiation at shorter wavelengths is not completely blocked but attenuated increasingly. For experiments it is important to know the transmission at each wavelength.

We often need filters which transmit a wavelength band only a few nm wide while they block all wavelengths outside this range. This property is found in interference filters which operate as Fabry–Perot interferometers. They are manufactured from several alternating transparent layers with different

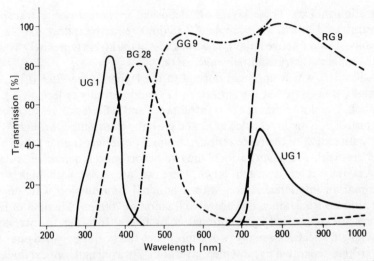

Fig. 3.5. Transmission (ordinate, percentages) of different band filters in dependence of the wavelength (abscissa in nm) (from Schott & Gen., modified).

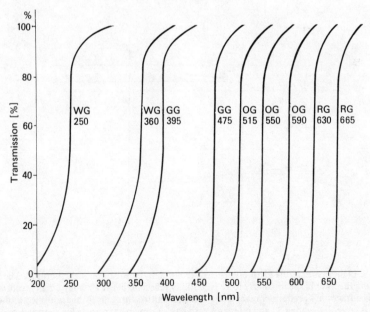

Fig. 3.6. Transmission (ordinate, in per cent) of different cut-off filters in dependence of the wavelength (abscissa, in nm) (from Schott & Gen., modified).

refraction indices. These layers are deposited in vacuum on a transparent substrate and consist for example of ZnS (high refractive index) and Na_3AlF_6 (cryolite, low refractive index). The thickness of the layers is precisely a quarter of the maximally transmitted wavelength.

A beam I_0 which impinges normal to the layers of the filter (Fig. 3.7) is partially transmitted at the surface F_2 (T_1) and partially reflected (ρ_1). (The symbols T and ρ represent the transmittance and reflectance, respectively). A fraction of ρ_1 is again reflected at F_1 and partially transmitted through F_2 (T_2). The path length of the twice reflected beam T_2 is longer than T_1 (by $2a$) and interferes with T_1. When $2a = n2\lambda$ the two beams cancel each other and when $2a = n\lambda$ they intensity each other. Here we are faced with higher-order maxima: an interference filter with a nominal transmission wavelength of 350 nm also transmits a considerable fraction at 700 nm. Maxima of higher orders are suppressed by additional color filters. In order to increase the wavelength selection, usually two or more such filters are deposited one on top of the other separated by 'absentee' layers of 1/2-wavelength optical thickness.

When the beam strikes the filter at an oblique angle the distance $2a$ is increased, and therefore the maximally transmitted wavelength is shifted to shorter wavelengths

$$\lambda_w = \lambda_s \frac{\sqrt{n^2 - \sin^2 \alpha}}{n}$$

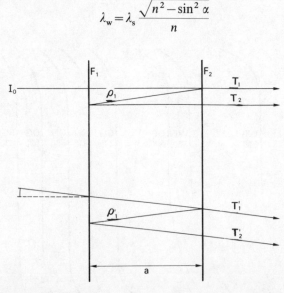

Fig. 3.7. Light rays entering an interference filter normally (top) or obliquely (bottom). The beam I_0 hitting the filter perpendicularly to F_1 is partially reflected (ρ_1) at F_2 and partially transmitted (T_1). The fraction T_2 reflected again at F_1 and transmitted through F_2 has travelled a distance which is larger by $2a$ than T_1 and interferes with it $(\rho_1$ is drawn at an angle for clarity). For beams deviating from the normal by an angle α the distance $2a$ increases, and the maximally transmitted wavelength is shifted to shorter wavelengths.

where λ_w is the wavelength transmitted when the beam is incident at an angle α and λ_s is the transmitted wavelength for perpendicular incidence and n is the refractive index of the layer. This phenomenon can be used to shift the transmission wavelength of the filter up to 10 nm towards shorter wavelengths without changing the transmission curve noticeable. Larger wavelength shifts are not possible without decreasing the maximal transmission.

The optical characteristics of interference filters can be described by a number of parameters (Fig. 3.8). The transmittance at the maximal transmitted wavelength λ_{max} is defined as T_{max}. The commercially available interference filters have a T_{max} of about 40%. The halfband width describes the wavelength band in which $\geq 50\%$ of T_{max} is transmitted. We could equally well state a band width for 10% or for 1% of T_{max}. Double interference filters are produced with halfband widths of about 5 nm but the narrow transmittance curve is traded for a small maximal transmittance of about 10%. In all color filters we are interested in the leakage at wavelengths outside the transmitted band. As an example, if we want to measure the efficiency of red light on a blue light effect we use a red light transmitting 650 nm interference filter which has a blue light transmittance of 0.001. Radiation through this red filter will cause the same reaction as a blue light filter when we use a 1000 times higher fluence

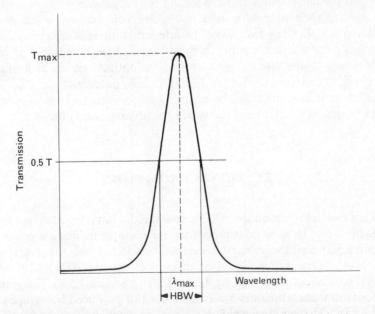

Fig. 3.8. Schematic transmission curve of an interference filter characterized by the maximal transmission T_{max} at λ_{max} and the half band width (HBW) at $0.5T$. In good-quality filters transmission outside the transmission band is about 0.1%.

rate than with the blue filter. This effect is, of course, not due to the red light sensitivity of the photoreceptor but to the leakage of the red filter at other wavelengths.

3.4.3   Polarization filters

In Section 1.2 we mentioned that radiation from a light source consists of waves with different polarization planes while linearly polarized light oscillates in one plane. A number of different materials can filter waves swinging preferentially in one plane out of a mixture of differently polarized waves (Fig. 3.9a). Polarized light is also produced by reflection (see Section 1.5.1) or by passing it through birefringent materials such as calcite. Some prism combinations also polarize light. Today polarized light is most easily produced by sheet-type dichroic polarizers made of polymeric plastic, the molecules of which are oriented by stretching the material mechanically in one direction. Two polarization filters in sequence only allow light to pass when their polarization planes are parallel to each other. Increasing the deviation angles reduces the transmission and crossed polarizers will extinguish nearly all radiation. This method can be used to control the radiation continuously, provided the polarization of the radiation is no handicap.

A special form of polarization is circular polarization, which can be produced by allowing two waves to interfere with each other which are polarized perpendicular to eath other and have a phase difference. The vector of this wave rotates along a helix around the optical axis. By reducing the amplitude of one wave we can produce elliptically polarized light (Fig. 3.9b). Thus linear polarization can be regarded as an extreme case of elliptical polzarization in which one of the waves has an amplitude of zero.

3.5   MONOCHROMATORS

Another method to produce monochromatic radiation is based on spreading white light into its spectral components, for example by using a prism (see Section 1.5.2). Small wavelength bands can be selected from the spectrum by slits. More often gratings are used in monochromators (Fig. 3.10) which spread the incident white light by diffraction (see Section 1.5.2). The gratings of modern monochromators are either etched or produced holographically (Fig. 3.11), which allows a small grating spacing and reduces the fraction of scattered light. Gratings are placed on a concave surface which is either spherical or–in better instruments–toroidal. As in interference filters (see

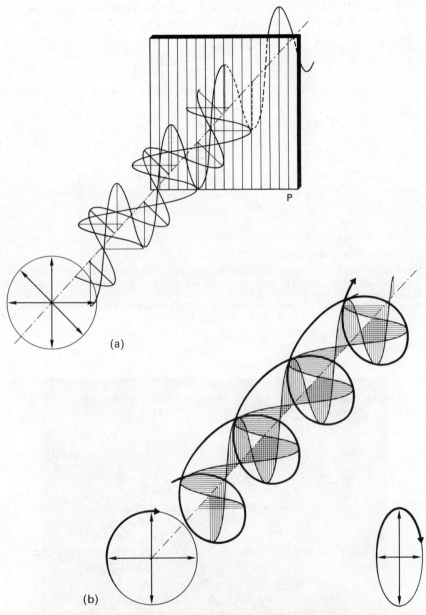

Fig. 3.9. (a) Polarization of unpolarized radiation by a polarization filter (P). The incident radiation is a mixture of waves oscillating in all possible planes (three of which are shown). The polarization filter transmits only the vertically oscillating component (from Gerlach, modified). (b) Circularly polarized light is produced by interference of two linearly polarized waves perpendicular to each other, which have a phase difference of 90°. By decreasing the amplitude of one wave we obtain elliptically polarized light (inset) (from Michel, modified).

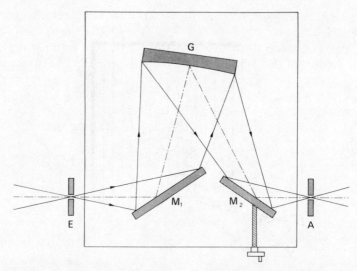

Fig. 3.10. Schematic diagram of a grating monochromator. The radiation entering through slit E is focussed onto the grating G by the mirror M_1. The radiation is spread by refraction and leaves the monochromator via mirror M_2 and the exit slit A. Wavelength selection is effected by tilting the mirror M_2 by means of a mechanical transmission (HL monochromator, courtesy of Jobin & Yvon).

Fig. 3.11. Scanning electron micrograph of a grating with 2400 lines per mm (original photograph courtesy of Jobin & Yvon).

Section 3.4.2), in monochromators we are faced with the problem of higher-order wavelengths.

The radiation enters the monochromator through an entrance slit E and falls on the mirror M_1 which reflects it onto the grating G. The selected monochromatic band leaves the instrument via mirror M_2 and exit slit A. The width of the wavelength band depends on the slit width. The desired wavelength is selected by changing the adjustment of the grating or a mirror by means of a mechanical transmission and can be read off a dial.

In order to produce an even sharper wavelength selection, the beam exiting from a first monochromator can be sent through a second one. The construction of such a double monochromator is shown in Fig. 3.12.

Fig. 3.12. Schematic diagram of the rays in a double monochromator (HRD, Czerny–Turner-type; original photograph courtesy of Jobin & Yvon).

Monochromators have an advantage over interference filters in that they transmit most of the incident energy of a wavelength band. This is because the gratings are coated with a reflecting metal layer. In addition, the wavelength can be adjusted over a wide range and the transmittance outside the selected range is small and is caused only by scattered light inside the monochromator.

3.6 BIBLIOGRAPHY

Textbooks and review articles
Beyer, H. (ed.) *Handbuch der Mikroskopie*. VEB Verlag Technik, Berlin (1973)
Emmerich, H. *Stoffwechselphysiologisches Praktikum*. Thieme, Stuttgart and New York (1980)
Gerlach, D. *Das Lichtmikroskop*. Thieme, Stuttgart (1976)
Halldal, P. (ed.) *Photobiology of Microorganisms*. Wiley Interscience, London, New York, Sidney and Toronto (1970)
Hartmann, K. M. Aktionsspektrometrie. In: Hoppe, W., Lohmann, W., Markl, H. and Ziegler, H. (eds) *Biophysik*, 2nd edn. Springer, Berlin and Heidelberg, pp. 197–222 (1982)
Hoppe, W., Lohmann, W., Markl, H. and Ziegler, H. (eds) *Biophysik*, 2nd edn. Springer, Berlin, Heidelberg and New York (1982)
Michel, K. *Die Grundzüge der Theorie des Mikroskops*. Wiss. Verlagsges, Stuttgart (1964)
Mohr, H. and Schopfer, P. *Lehrbuch der Pflanzenphysiologie*. Springer, Berlin, Heidelberg and New York (1978)
Schäfer, E. and Fukshansky, L. Action spectroscopy. In: Smith, H. and Holmes, M. G. (eds) *Techniques in Photomorphogenesis*. Academic Press, London, pp. 109–129 (1984)
Swanson, C. P. (ed.) *An Introduction to Photobiology*. Prentice-Hall, London (1969)
Urbach, W., Rupp, W. and Sturm, H. *Experimente zur Stoffwechselphysiologie der Pflanzen*. Thieme, Stuttgart (1976)

Further reading
Arditti, J. and Dunn, A. *Experimental Plant Physiology*. Holt, Rinehart and Winston, New York (1969)
Cabana, A., Doucet, Y., Garneau, J.-M., Pepin, C. and Puget, P. The vibration–rotation spectrum of methinophosphide: the overtone bands $2v_1$ and $2v_3$, the summation bands $v_1 + v_2$ and $v_2 + v_3$, and the difference band $v_1 - v_2$. *J. Mol. Spectr.* **96**, 342–350 (1982)
Duckrow, R. B., LaManna, J. C. and Rosenthal, M. Sensitive and inexpensive dual-wavelength reflection spectrophotometry using interference filters. *Anal. Biochem.* **125**, 13–23 (1982)
French, C. S. Sharper action spectra. *Photochem. Photobiol.* **25**, 159–169 (1977)
Hartmann, K. M. and Cohen-Unser, I. Analytical action spectroscopy with living systems: photochemical aspects and attenuance. *Ber. Deutsch. Bot. Ges.* **85**, 481–551 (1972)
Jobin & Yvon. *Light and Physics*. Imprex, no publ. date
Lee, Y. J. and Burr, J. G. Singlet and triplet energy transfer in α-diketone derivates or uracil and thymine. *Photochem. Photobiol.* **37**, 381–389 (1983)
Melles Griot. *Optics Guide*. Arnhem, Holland (1984)
The Merck Index. 7th edn. Rahway, New Jersey (1960)
Oriel Corporation of America, Catalog
Ray, P. M. *The Living Plant*. Holt, Rinehart and Winston, New York (1963)
Rolyn Optics Company: *Optics for Industry*. Catalog 184 (1984)
Schott & Gen. *Farb- und Filterglas*, Mainz, o.J.
Shibata, K. Spectrophotometry of intact biological materials. *J. Biochem.* **45**, 599 (1958)
Shropshire, W. Action spectroscopy. In: Mitrakos, K. and Shropshire, W. (eds.) *Phytochrome*. Academic Press, New York, pp. 161–181 (1972).

4 Measurement of radiation

The measurement of radiation energy is based on a comparison with a standard, regardless of whether we measure in the energetic or photometric systems (see Section 1.4). In a simple arrangement a standard observer (whose spectral sensitivity agrees with the V_λ curve which was standardized in 1951 by the CIE in Stockholm) compares the brightness of a standard light source with that of a measured source. Keeping the distance to the standard light source constant, the observer adjusts both light sources to the same brightness by varying the distance to the measured light source. From this he can determine the illuminance of the measured light source, since the illuminance decreases with the distance squared.

Standard light sources are available, for instance, at the National Bureau of Standards (Washington), the Laboratoire Central d'Electricité (Paris) or the Physikalisch–Technische Bundesanstalt (Braunschweig), against which secondary standard sources are calibrated; from the latter, standard sources are derived for practical use.

4.1 SELECTIVE RADIATION DETECTORS

Modern optoelectronic radiation detectors can be defined as devices which change their electrical properties, such as current, voltage, resistance or conductivity under the influence of light. Since we are interested in a linear relationship between fluence rate and the measured electrical parameters we can exclude from our considerations devices, such as photothyristor or photo-Schmitt-trigger, which change their properties at a critical threshold. Optoelectronic detectors can be divided into devices which utilize either the outer or the inner photoeffect.

4.1.1 Outer photoeffect

The outer photoeffect is characterized by the fact that electrons are emitted from a conducting material into the surrounding vacuum or rare gas when hit

GP–C

47

by quanta (Einstein, in 1905). All devices using the outer photoeffect are pronounced as tubes such as photocells and photomultipliers, as well as some television camera tubes.

A photocell is manufactured from an evacuated or gas-filled tube (Fig. 4.1). The radiation passes through a window in the metal-coated inside of the tube which forms the cathode; the cathode emits electrons under irradiation. An externally applied voltage accelerates the electrons towards the anode. The photocurrent is proportional to the photon fluence rate within certain limits. In gas-filled photocells the electrons collide with gas molecules, which in turn emit electrons, so that the electron current is amplified. All photocells show a dark current which is due to the finite isolation resistance between cathode and anode. In addition, at room temperature some electrons leave the cathode even in darkness. During measurement the photocurrent should be at least 100 times the dark current, to keep the error low.

Each photocell has its own spectral sensitivity. Cathodes made from copper, gold or silver emit electrons only when struck by ultraviolet radiation. Alkali metals such as lithium, sodium, potassium, rubidium and cesium have an upper limit in the visible or infrared range. Photocells are used in the laboratory for standard measurements. Their usage, however, is limited by their low sensitivity and their wavelength-dependence.

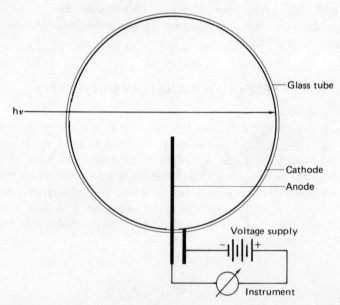

Fig. 4.1. Schematic diagram of a photocell. The radiation enters the photocell through a window in the metallic coating on the inside of the tube, which acts as the cathode and emits electrons when irradiated. The electrons are attracted by the positively charged anode. The photocurrent can be read off a meter.

Extremely low radiation levels can be measured using a photomultiplier which utilizes the amplification effect of secondary electrons. The electrons emitted from the cathode are attracted to an auxiliary anode (dynode) with a more positive voltage. The dynode emits five to ten secondary electrons when struck by the first electron. The amplified electron current strikes a second dynode with a higher positive electrical potential than the first one, and the amplification process is repeated. In commercially available photomultipliers six to thirteen dynodes are mounted in sequence (Fig. 4.2), which increases the amplification accordingly. Depending on the application (e.g. high sensitivity, frequency or photocurrent), a resistor cascade is selected which defines the potentials of the various dynodes.

Photomultiplier tubes can detect extremely low photon fluence rates or even single photons. However, we have to take into account the dark current, most of which is due to thermal effects. Therefore the dark current can be drastically decreased by cooling the photomultiplier in, for instance, liquid nitrogen. Because of the high amplification the photocurrent is very dependent on the stability of the supply voltage which should be kept constant to within

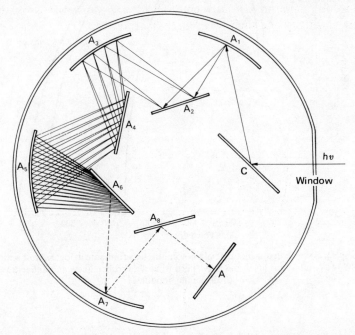

Fig. 4.2. Schematic diagram of a photomultiplier. The radiation enters through a window and hits the photocathode (C) which emits electrons. The electrons are attracted by the first dynode (A_1) where they liberate several secondary electrons. This process is repeated at each subsequent dynode (A_2 through A_8), which have increasingly higher positive potentials, and finally reach the anode (A). The amplified photocurrent can be measured (from Greif, 1972, modified).

±0.1% at the most. Photomultipliers are usually used for small fluence rates, since a high current between the last dynode and the anode, due to the high amplification, can overheat the device.

Photomultipliers show a strong dependence of the incident wavelength (Fig. 4.3), as do photocells. Depending on the application we select a cathode material which emits maximally when irradiated in the ultraviolet or visible range. The window has to be selected in its transmission as well. It is advisable to cool the photomultiplier, especially when used in the red range, since the dark current increases at longer wavelengths.

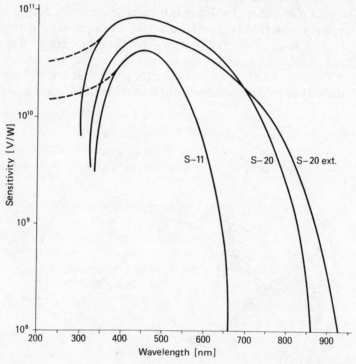

Fig. 4.3. Spectral sensitivity of commercial photomultipliers (ordinate in logarithmic scale) using glass windows (continuous lines) and fused silica windows (broken lines) and different cathodes (from Ealing, modified).

4.1.2   Inner photoeffect

All devices which make use of the inner photoeffect consist of an electrically semiconducting material. Electrons released upon irradiation do not leave the crystal but increase the electric conductivity due to their free mobility. We can distinguish photosemiconductors with and without a barrier layer.

Photoresistors are manufactured from an unpolarized semiconducting material which changes its conductivity upon irradiation. Among the many substances known to show an inner photoeffect, CdS and PbS are most commonly used. In addition, selenides, arsenides and antimonides are employed. The material coats an insulating material, such as glass, and the sandwich is sealed in a gas-tight but transparent case.

In photoresistors the dependency between the inner resistance and the fluence rate follows a steep curve. They need an external voltage supply to produce a measurable current. The spectral sensitivity of photoresistors made from CdS resembles that of the human eye, which excludes measurements in the infrared and ultraviolet. Other disadvantages are their temperature–dependency and their often considerable hysteresis: the curves differ depending whether they have been measured using increasing or decreasing fluence rates. The history of a photoresistor is important for its characteristics, and high fluence rates can decrease its efficiency by up to 20%. In addition, photoresistors of the same type often differ considerably in their electrical characteristics, so that they have to be selected for similar behavior when they are used in a Wheatstone bridge. Due to the many problems described, photoresistors are usually only used in measurements which do not require high precision.

Photodiodes and phototransistors are polarized semiconductors which use the inner photoeffect at a barrier layer (Fig. 4.4). A photodiode is made of a crystal of semiconducting silicium or germanium. One half has been made a p-type semiconductor by adding acceptor atoms such as boron or gold to the original pure crystal (p-region). The other one is transformed into an n-type semiconductor (n-region) by adding electron donor atoms such as antimony,

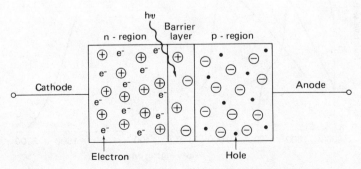

Fig. 4.4. Schematic diagram of a semiconducting crystal in a photodiode. For details see text (from Jahn).

arsenic, or phosphorus. Due to their thermal migrations in darkness, electrons move into the p-region and holes (positive charges) into the n-region where they recombine with the acceptor and donor atoms located as impurities in the crystal lattice. These charge migrations result in a barrier layer free of mobile charges. Near the barrier layer the charge separation induces an electrical field which, however, cannot be detected at the electrodes. Irradiation of the barrier layer induces the production of addition free electrons and holes which are accelerated in the created field. This photocurrent can be measured when the circuit is closed with an ammeter. In this configuration the photodiode operates as an photovoltaic element which produces a photocurrent without an externally applied voltage. Therefore this device can be easily used for outdoor measurements. The resistance and capacitance of the circuit define the time constant and thus the upper cutting-off frequency. The dark current can usually be neglected. The photocurrent depends on the wavelength of the incident radiation (Fig. 4.5) which can be described by the spectral quantum yield $\Phi(\lambda)$. The description of the physical mechanism indicates that the electric response–like that in photocells–does not depend on the energy fluence rate but on the number of quanta incident on an area per time unit. In practice one has to bear in mind that linearity holds only within certain limits.

In addition to their use as active photoelements, photodiodes can be used as photoconductive elements. In this case they operate as polarized light-dependent resistors in the nonconducting direction. Since their dark conductivity is quite low, silicium diodes can be used to detect low fluence

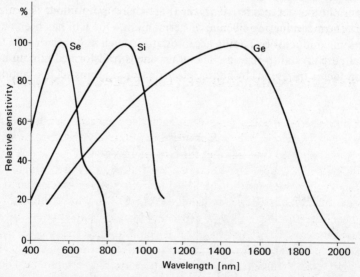

Fig. 4.5. Relative spectral sensitivity of Se, Si and Ge as crystal material in photodiodes as percentages of the maximal value (from Bergtold, modified).

rates. Germanium diodes, however, have a considerable dark current and a lower sensitivity in the visible range so that they can only be used at illuminances of greater than 100 lx.

PIN diodes represent a further development, and have an additional intrinsic region between the positive and negative regions which is either weakly positive or negative. The spectral sensitivity of PIN diodes extends up to 1000 nm. They allow the measurement of very small light fields since their active area is in the order of 10^{-2} to 10^{-3} cm^2.

In principle, photoelements are very large photodiodes made from selenium or silicium. The spectral sensitivity of selenium photoelements has a maximum between 550 and 590 nm. Therefore they are used in luxmeters since they can be easily corrected to fit the spectral sensitivity of the human eye. The photovoltage measured without an external load increases with the logarithm of the illuminance. For the practical application as a measurement device, temperature-dependence and aging need to be considered.

Si photoelements are also used as solar cells to convert light energy into electric energy, for example in space or in field applications independent of an electric supply.

Phototransistors use the same principle as normal (planar) transistors. They have a higher sensitivity than photodiodes (about 0.1 to 1 A/lm) which is due to the amplification factor β of the transistor, which is usually between 10 and 100. The amplification also increases the dark current by the same factor. Since amplification depends on fluence rate there is no linear dependence between photocurrent and fluence rate.

4.2 RADIATION MEASUREMENT BY ENERGY CONVERSION

Many photoelectric devices have a characteristic spectral sensitivity. This problem is overcome when the radiation is quantitatively absorbed by a black surface and converted into another form of energy, heat.

A bolometer measures the electrical resistance of a wire, which varies depending on the temperature of the wire. The wire is black-coated and absorption of radiation increases its temperature. The wire is mounted in a vacuum inside a glass tube to avoid any heat loss to the surrounding air. Two identical elements are combined in a bridge circuit: one is irradiated and the other is kept in darkness. This configuration makes the device independent of the ambient temperature. However, bolometers are not very practical due to their size, and do not allow measurements of small light fields.

Radiometers often employ a thermistor, which is a temperature-sensitive resistor made from a semiconducting crystal. This device also converts light into electric energy using absorption by a blackbody.

A thermopile consists of a number of thermocouples connected in sequence, each made of two different metals in electrical contact. A Moll's thermopile is built from strips of a constantan–manganese alloy connected by silver solder and rolled into thin bands. Every second junction is kept in darkness (reference junctions) while the radiation strikes the intermediate test junctions, which have been blackened. This arrangement induces a temperature difference which causes a measurable thermal current. The voltage of each connection depends on the metals used and amounts, for example, to 53 μV/degree centigrade for iron–constantan thermocouples.

Modern thermopiles are also very sensitive in the infrared region: the Epply Labs thermopile (12 bismuth–silver thermocouples) can detect the energy emitted from a cup of hot coffee at a distance of 1 meter. Both radiometers and thermopiles are often used in the laboratory, because they respond linearly over a wide range of fluence rates. However, they are not as sensitive as photomultipliers. Thermopiles measure energy fluence rates which can be converted into quantum fluence rates.

4.3   QUANTUM COUNTERS AND ACTINOMETERS

The problem of spectral sensitivity of photodectors can be solved using quantum counters, which are also often used in the UV range. The term indicates that these devices measure incident quantum fluence rates independent of the wavelength. The radiation to be measured is absorbed by an optically dense suspension of a fluorescent substance mounted in front of a photodetector. Substances are selected which show a constant fluorescence quantum yield over a wide wavelength range. The quantum yield of rhodamine B (8 g/l dissolved in ethylene glycol) only varies by $\pm 11\%$ in the range between 250 and 600 nm and by $\pm 1.5\%$ in the range between 400 and 600 nm. Thus, while the excitation wavelength can vary over a wide range, the emission wavelength is constant. Therefore the device is independent of the spectral sensitivity of the photodetector used.

Actinometers are often used to calibrate monochromators by measuring the quantum flux $I_0(\lambda)$ or the quantum dose $I_0(\lambda) \times \Delta t$ at a given wavelength. A chemical actinometer is a substance which undergoes a chemical reaction, such as a self-sensitized photooxidation, upon irradiation. The total photon fluence can be determined afterwards, for example spectrophotometrically or by titration. The quantum yield should be thermally stable and independent of the incident wavelength in the specified band. If the substance shows an absorbance change at a measurement wavelength when irradiated at a test wavelength λ the quantum yield $\Phi(\lambda)$ at the irradiation wavelength λ is calculated from

$$\Phi(\lambda) = \frac{\Delta E V}{E l I_0(\lambda) A(\lambda) \, \Delta t}$$

where E is the optical density and ΔE the change in the optical density at the measurement wavelength during the time interval Δt, $A(\lambda)$ the mean absorption at λ and l the pathlength in the cuvette.

One of the classical substances is uranyl oxalate: the amount of oxalate still present after irradiation is determined. The uranyl oxalate method is relatively insensitive and therefore long exposure times (minutes to hours) are necessary. Recently, more sensitive substances have been found, such as potassium ferrioxalate, chromium complexes and heterocoerdianthrone-endoperoxide.

4.4 SPECTROPHOTOMETERS

In order to automatically record absorption spectra, spectrophotometers have been developed, which essentially consist of the elements described above (Fig. 4.6). The measurement beam from a light source is spread into its spectral components by a monochromator (or double monochromator) and passed through a variable slit and through the measured solution in a cuvette onto a photomultiplier. In order to cover a wider spectral range, two selectable light sources are usually built in, one of which is used in the visible range (for example a tungsten iodine lamp) and the other in the ultraviolet range (for example a deuterium lamp) (see Section 2.2.2.1).

By scanning a certain wavelength range the photocurrent changes with the spectral absorption $E(\lambda)$ of the sample according to the Lambert Beer law (see Section 3.1). When we record the electronically amplified photocurrent during the scan on an XY recorder we obtain an absorption spectrum.

This spectrum depends on the emission spectrum of the light source and on the spectral sensitivity of the photodetector. In addition, all optical properties of the cuvette and the solvent affect the absorption spectrum. In order to eliminate all these problems the measurement beam is split into two beams, one of which passes through the sample while the other passes through a reference which is identical with the sample and only lacks the substance to be measured. A chopper sends the beam through the sample and the reference in rapid alternation. The signal is detected by one or two photomultipliers. The electronically calculated difference between sample and reference represents the absorption of the measured substance.

Using a spectrophotometer one has to be aware of some potential errors. The dark current of the photomultiplier determines the minimal photocurrent

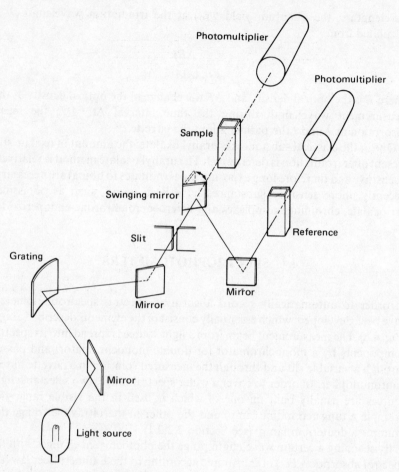

Fig. 4.6. Schematic diagram of a double beam spectrophotometer. For details see text.

(see Section 4.1.1). This and the scattered light limits the highest measurable absorbance; in commercial instruments the highest measureable absorbance is found at about 3 to 4. The old term 'optical density' (1 OD equals an absorbance of 1) should be avoided. The other problem is that the fluence rate of the light source cannot be increased infinitely. In addition to technical problems with generating high fluence rates, the measured substance could change its absorption characteristics under the influence of the measurement beam.

Scattering samples pose another problem: since the cuvette and the photomultiplier are mounted at a distance from each other only a fraction of the radiation scattered in all directions strikes the detector (Fig 4.7) while all of

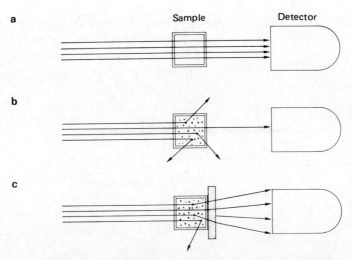

Fig. 4.7. Shibata method to reduce problems caused by scattering samples. All radiation not absorbed hits the detector when the sample shows no scattering (a). In a strongly scattering sample the non-scattered fraction hits the detector while most of the scattered fraction misses it (b). By inserting a frosted glass (c) the total radiation in the direction of the detector is scattered, which reduces the fraction of the unscattered radiation considerably (from Shibata, 1959).

the unscattered radiation is caught by the detector. This effect is even more marked at shorter wavelengths than at longer ones (see Section 3.2), which feigns a high absorption of shorter wavelengths: the absorption spectrum is superimposed by an exponential function increasing toward the blue region. Shibata (1959) has described a method which reduces this problem. The radiation leaving the sample is scattered by opalescent glass. This technique reduces the fraction of the unscattered radiation and consequently shoulders and maxima are more distinctly visible. Another method is to equalize the amount of scattering in both the sample and reference by adding, for example, 1 g $CaCO_3$ per ml. The scattering at the $CaCO_3$ particles should exceed that of the measured suspension so that a large fraction of scattering is eliminated by taking the difference between sample and reference.

The technique of differential spectrophotometry allows detection of very small absorption differences. When we insert the same substance in the sample and in the reference beam at the same concentration, and illuminate the sample while we keep the reference in darkness, the difference spectrum indicates light-induced absorption changes. Similarly, we can determine absorption differences between oxidized and reduced substances.

The bandwidth of an absorption band depends on a number of factors, including the temperature. The absorption of a quantum raises the molecule from its ground state into an excited state. The wavelength of the

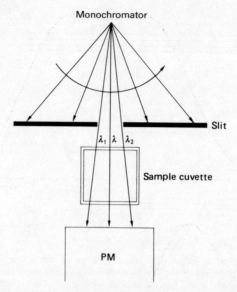

Fig. 4.8. Artifactual broadening of absorption peaks in a spectrophotometer. The photomultiplier PM detects not only the selected wavelength λ, but the whole spectral range from λ_1 to λ_2 transmitted through the slit.

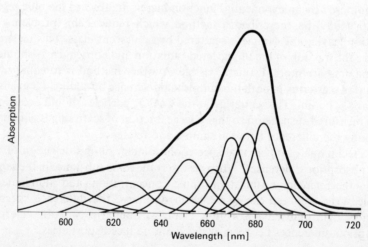

Fig. 4.9. Deconvolution of an absorption spectrum of spinach chloroplasts. The mathematical analysis provides the absorption spectra of the postulated individual pigments. The superimposition of all individual absorption spectra yields the total absorption spectrum which corresponds with the action spectrum (from van Ginkel and Hammans, 1980).

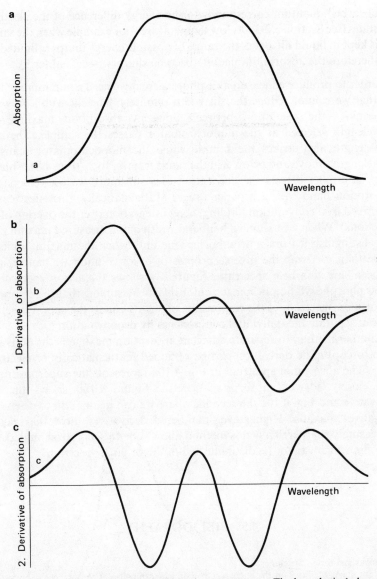

Fig. 4.10. Schematic diagram explaining derivative spectroscopy. The hypothetical absorption spectrum (a) (with two adjacent maxima) and the first (b) and second derivative (c). The two (negative) maxima in the second derivative are clearly separated and their peaks correspond with the peaks in the original spectrum.

absorbed quantum corresponds to the energy difference of the electronic states (see Section 5.2.3). At low temperatures–for example when the sample is kept in liquid nitrogen–the range of possible energy jumps is limited and therefore the absorption maxima become sharper (see Chapter 9).

In order to produce a measureable photocurrent we need a minimum of light, so that we cannot reduce the slit width infinitely. The slit width, however, determines the resolution between adjacent absorption maxima. The wavelength selected at the monochromator passes the sample. The same wavelength also strikes the sample while the monochromator scans the wavelengths above and below λ in the range from λ_1 to λ_2 (Fig. 4.8). Thus the photomultiplier detects the absorption of a substance with a very narrow absorption band over a broader range. Mathematically, this effect can be described by a convolution (folding, here: integration over the overlap of two functions). When we assume a halfband width a of a measured maximum we can distinguish it from a neighboring one only when the maxima differ by more than a. Using the inverse process of deconvolution we can separate adjacent maxima in a spectrum. Figure 4.9 shows the action spectrum of photophosphorylation in spinach chloroplasts. Assuming that the composite absorption spectrum of the responsible pigments equals the action spectrum we can separate the individual components by deconvolution.

Another method to separate adjacent absorption maxima is the derivative spectroscopy. The derivation can be obtained mathematically or electronically. The absorption spectrum in Fig. 4.10a represents the superposition of two closely adjacent absorption spectra. Figure 4.10b shows the first derivative and Fig. 4.10c the second, where we can distinguish two separate (negative) maxima. Higher even-numbered derivatives show the maxima increasingly sharper. Thus this method allows one to separate shoulders and maxima which cannot be distinguished in the original spectrum.

4.5 BIBLIOGRAPHY

Textbooks and review articles
Benedetti, P. A. Methods for the microspectroscopic investigation of photoreceptive structures. In: Colombetti, G., Lenci, F. and Song, P.-S. (eds). *Sensory Perception and Transduction in Aneural Organisms*. Plenum Press, New York and London, pp. 61–74 (1985)
Bergmann, L. and Schaefer, C. Lehrbuch der Experimentalphysik. III. In: Matossi, F. (ed.) *Optik*, 4. Aufl. Walter de Gruyter, Berlin (1966)
Demas, J. N. and Crosby, G. A. The measurement of photoluminescence quantum yields. A review. *J. Phys. Chem.* **75**, 991–1024 (1971)
Gerthsen, C. and Kneser, H. H. *Physik*. Springer, Berlin Heidelberg and New York (1971)
Greif, H. *Lichtelektrische Empfänger*. Akademische Verlagsgesellschaft, Leipzig (1972)
Gross, H., Seyfried, M., Fukshansky, L. and Schäfer, E. *In vivo* spectrophotometry. In: Smith, H.

and Holmes, M. G. (eds) *Techniques in Photomorphogenesis.* Academic Press, London, pp. 131–157 (1984)

Holmes, M. G. Radiation measurement. In: Smith, H. and Holmes, M. G. (eds.) *Techniques in Photomorphogenesis.* Academic Press, London, pp. 81–107 (1984)

Keitz, H. *Light Calculations and Measurement.* Macmillan, London (1971)

Reeb, O. *Grundlagen der Photometrie* (1962)

Wehrey, E. L. Molecular fluorescence, phosphorescence and chemiluminescence spectrometry. *Anal. Chem.* **54**, 131R–150R (1982)

Further reading

Bergtold, F. Lichtfühler–Spektren. *Funkschau* **42**, 134 (1970)

Brauer, H.-D. and Schmidt, R. A new reusable chemical actinometer for UV irradiation in the 248–334 nm range. *Photochem. Photobiol.* **37**, 587–591 (1983)

Frye, S. L., Ko, J. and Halpeern, A. M. Improved analysis of time-resolved single photon counting fluorescence data of excimer/exciple systems using a dual analysis iterated reconvolution program on a dedicated laboratory microcomputer. *Photochem. Photobiol.* **40**, 555–561 (1984)

Funktechnische Arbeitsblätter. Die PIN-Diode und ihre Anwendung. *Funkschau* **45**, 37 (1973)

van Ginkel, G. and Hammans, J. W. K. Action spectra of photophosphorylation. II. ATP formation catalyzed by phenazinemethosulphate suggesting the involvement of long wavelength pigment forms in the light harvesting process from PS II and PS I. *Photochem. Photobiol.* **31**, 385–395 (1980)

Griffith, O. H., Houle, W. A., Kongslie, K. F. and Sukow, W. W. Photoelectron microscopy and photoelectron quantum yields of the fluorescent dyes fluorescein and rhodamine. *Ultramicroscopy* **12**, 299–308 (1984)

Hsiao, K. C. and Björn, F. O. An artifact in measurements of *in vivo* light-induced absorbance changes. *J. Biochem. Biophys. Methods* **8**, 271–274 (1983)

Jahn, G. Physikalische Grundlagen der Optoelektronik. *Funkschau* **49**, I, 55; II, 71; III, 163 (1977)

Köhler, A. and Schiffel, R. Solarzelle, Solargenerator. Teil 1. *Funkschau* **23**, 65–66 (1985)

Köhler, A. and Schiffel, R. Solarzelle, Solargenerator. Teil 2. *Funkschau* **24**, 63–66 (1985)

RCA Electro-Optics Handbook. Lancaster (1974)

Reiher, D. Die Fertigung integrierter Schaltungen. *Funkschau* **40**, I, 591; II, 671 (1968)

Schiffel, R. Sender und Empfänger für die Lichtleitertechnik. *Funkschau* **13**, 55–61 (1985)

Schleicher, A. and Hofmann, K. P. Time-resolved angular dependent measurement of triggered light scattering changes in biological suspensions. *J. Biochem. Biophys. Methods* **8**, 227–237 (1983)

Shibata, K. Spectrophotometry of translucent biological materials. Opal glass transmission method. *Methods Biochem. Anal.* **7**, 77–109 (1959)

Texas Instruments Inc. *The Optoelectronics Data Book.* Dallas (no publ. date)

PART II

General Photochemistry

5   Excited states and their relaxation

Each photochemical process begins with the excitation of an electron: the energy of the electromagnetic radiation is absorbed by an electron which is lifted to a higher energy level. Electrons of both atoms and molecules can be excited. Their return to the ground state is called relaxation.

5.1   EXCITATION OF ATOMS

We will begin by discussing the simple case of atoms in a gaseous phase in such a low concentration that the atoms do not influence each other. The electrons are in the ground state, the energetically lowest possible state. By absorbing energy they can enter an excited state. One important limitation is that each electron – and thus each atom – can only occupy distinct energy levels; intermediate levels are 'forbidden'. Since quanta cannot be divided an electron can only be excited by such quanta, the energy of which corresponds with the energy difference from one energy level to a higher one. These energy levels can be symbolized by the rungs on a ladder, though the distance between the rungs is not uniform. The energy of a quantum can be stated in electron volts (eV); it is proportional to the frequency and inversely proportional to the wavelength of the electromagnetic radiation (see Section 1.3). Therefore, from a broad spectrum only that wavelength is absorbed, the energy of which corresponds to the excitation energy of the electron. When the electron returns from the excited state to the ground state it releases this energy difference, for example by emitting a photon at the corresponding wavelength.

The electrons occupy orbitals which can be described by a wave function. One can symbolize the probability of occupancy of an electron in three-dimensional space by charge clouds whose boundaries are defined by the fact that they include a certain amount (for instance 95%) of the negative charge. Each orbital can be occupied by no more than two electrons. An electron can be regarded as a negatively charged particle which, due to its rotation, generates a circular magnetic field. The rotation around a central axis is defined as the spin of the electron. In the ground state the spins of the two paired electrons are antiparallel (opposite) according to the Pauli principle.

This spin combination has only one configuration – the spin of one electron is in one direction and that of the other in the opposite direction – which can be calculated using the multiplicity M: when we define the spins of the electrons as $-1/2$ and $+1/2$ we obtain

$$M = 2|S| + 1 = 1$$

since the sum of the two spins is $S = +1/2 - 1/2 = 0$. Therefore this state of an atom or molecule is called the singlet state. The same relationship holds for unpaired electrons with antiparallel spins which can result from the fact that one electron has been excited to a higher state. If one of the electrons inverts its spin the multiplicity calculates to

$$M = 2|S| + 1 = 3$$

since the sum of the two spins $S = \pm 1/2 + \pm 1/2 = \pm 1$. This configuration is called the triplet state where the sum of the spin vectors can be either -1 or $+1$. In fact we find three different spin isomers which can be distinguished by their different magnetic momentum when subjected to a strong magnetic field. The two electrons which occupy a common low-energy orbital are in such a stable configuration that a large amount of energy is necessary to lift one of the electrons to an energetically higher free orbital. Usually the energy of visible light is not sufficient, and it takes ultraviolet radiation or even X-rays to do that. In addition to paired electrons, an atom can have unpaired electrons in the outer orbitals. When these electrons occupy an s-orbital they are called s-electrons and when they occupy a p-orbital they are called p-electrons. These electrons require less energy to be excited to a higher orbital.

Ten of the eleven electrons of the sodium atom occupy filled orbitals as n-electrons; the only remaining one is an s-electron. In the ground state it occupies an orbital described as $3S_{1/2}$. The index $1/2$ indicates the spin of the electron. The next higher rung in the energy ladder is the orbital $3P_{1/2}$ (Fig. 5.1). The energy difference amounts to 2.103 eV, which corresponds to a wavelength of 589.6 nm. The transition from the ground state to the first excited singlet state can formally be written as $S_0 \rightarrow S_1$. When a quantum at a wavelength of 589.1 nm is absorbed, which corresponds to an energy of 2.105 eV, the electron enters the orbital $3P_{3/2}$ and changes its spin. In this case the atom makes a transition to the first excited triplet state T_1.

Above the first excited singlet and triplet states there is a sequence of rungs with increasingly closer-spaced energy levels. On each higher level the electron attains a greater distance from the nucleus and approaches the ionization limit. At this point the electron has absorbed sufficient energy to be separated as a free electron from the nucleus and to leave the atom as a positively charged ion. For the sodium atom this energy amounts to 5.12 eV. During recombination of ion and electron this energy is usually not released in one

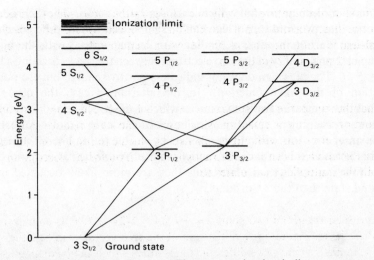

Fig. 5.1. Energy diagram of the sodium atom. The atom can absorb only discrete energy portions (which correspond to certain wavelengths). the electron is activated to the next higher excited state. During relaxation the atom emits a quantum with the corresponding wavelength when it returns to a lower state (from Nassau, 1980).

step, but the electron passes through several intermediate states accompanied by the emission of low-energy quanta during each step. This radiation represents the characteristic monochromatic emission lines of sodium light. Similar lines are emitted by mercury or neon lamps or by gas lasers (see Section 2.2.2) though at different wavelengths due to the different energy levels of the orbitals in these atoms.

5.2   ABSORPTION BY MOLECULES

The unpaired s- and p-electrons in the outermost orbitals can share a common orbital with the electrons of other atoms and therefore bond atoms together to form molecules (valence electrons). In the binding of atoms different electrons can be involved and different new orbitals can be formed. In the following we will characterize these orbitals before we disuss the excitation of electrons in these orbitals.

5.2.1   Molecular orbitals

In the hydrogen molecule the two s-electrons, which are contributed by the two participating hydrogen atoms, are called σ-electrons and occupy a common molecular orbital which encloses both nuclei. The Schrödinger equation has two solutions which can be symbolized by two different charge densities. We obtain either a binding σ-molecular orbital with the highest occupancy probability of the two electrons between the nuclei forming a bond (Fig. 5.2). (The plus sign does not indicate a positive charge but the positive fraction of the wave function.) In the alternative case the occupancy probability is exterior to the two nuclei which is described as a σ^*-orbital and causes a weakening or breaking of the bond. Electrons in σ-molecular orbitals have an energy too weak to be excited by visible light. The σ^*-molecular orbital levels are also above the energy range of visible light. Since σ-orbitals have lower energies than σ^*-orbitals they are more likely occupied in the ground state than the σ^*-orbitals.

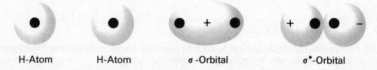

H-Atom   H-Atom σ-Orbital σ^*-Orbital

Fig. 5.2. Formation of molecular orbitals from the atomic orbitals of two hydrogen atoms. According to the two solutions of the Schrödinger equation there is a binding σ-molecular orbital and a non binding σ^*-molecular orbital (from Christen, modified).

The n-molecular orbitals do not form bonds since they do not concentrate enough electron density between the nuclei. In double bonds (such as $C = C$ or $C = O$ bonds), we have π-molecular orbitals in addition to σ-orbitals (Fig. 5.4). They are formed by the overlap of orbitals for the two p-electrons of two adjacent atoms which contribute one electron each to the bond. According to the two solutions of the Schrödinger equation (see above), there are two possible energy levels: one for a binding π-orbital and one for an antibinding π^*-orbital. These two orbitals are photobiologically of great interest since they can be excited by visible light or by the adjacent infrared and ultraviolet radiation.

In complex molecules with conjugated double bonds – where single bonds and double bonds alternate, π- and π^*-orbital electrons can be delocalized over several atoms. In benzene, for example, all p-orbitals fuse to form a π-molecular orbital (Fig. 5.3). Especially in molecules with extended conjugated double bonds (see Chapter 9) the π-orbitals are so extended that the energy of visible light is sufficient to excite the π-orbital electrons. Therefore we find many vividly colored substances among the complicated organic molecules with conjugated double bonds often extending over several ring systems. Additional side-chains with electron acceptors or donors can intensify the absorption and therefore the color; they are called auxochromes.

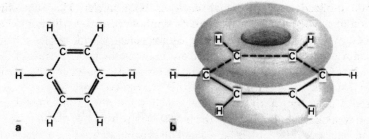

Fig. 5.3. (a) System of conjugated double bonds in benzene. (b) The π-orbitals fuse with one another (from Smith and Hanawalt, 1969, modified).

5.2.2   Excited states

After discussing the various molecular orbitals we will now schematically list all electron configurations of a molecule such as formaldehyde (CH_2O) in its ground state S_0. The exponents state the number of electrons in each orbital and the indices indicate the atoms involved.

$$S_0 = (1s_O)^2(1s_C)^2(2s_O)^2(\sigma_{CH})^2(\sigma_{CH'})^2(\sigma_{CO})^2(\pi_{CO})^2(n_O)^2(\pi_{CO}^*)^0(\sigma_{CO}^*)^0.$$

The bonding and non-binding orbitals are shown in Fig. 5.4.

When we consider only the orbitals which can be affected by visible or adjacent radiation we can exclude the σ- and the low-energy s-orbitals and we are left with the following orbitals for the singlet ground state (in descending order)

$$S_0 = (\pi_{CO})^2(n_O)^2(\pi_{CO}^*)^0(\sigma_{CO}^*)^0.$$

Various electronic transitions from a low-energy to a higher-energy molecular orbital can be induced by the absorption of radiation:

$$n \rightarrow \pi^*$$
$$\pi \rightarrow \pi^*$$
$$n \rightarrow \sigma^*$$
$$\pi \rightarrow \sigma^*$$

Due to the high energy it requires the $\sigma \rightarrow \sigma^*$ transition usually occurs in the vacuum UV region. An intense radiative absorption occurs during the $\pi \rightarrow \pi^*$ transition because of the large overlap between the wave functions in the ground state and the excited state. In contrast, the transition probability for $n \rightarrow \pi^*$ is 10 to 100 times lower because of the smaller overlap.

The statistical probability for an interaction between a photon and an absorbing molecule depends on a number of factors, including the time the electromagnetic wave spends near the molecule. At a wavelength of 550 nm a

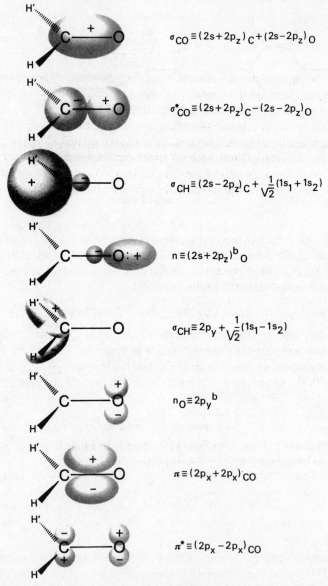

$$\sigma_{CO} \equiv (2s + 2p_z)_C + (2s - 2p_z)_O$$

$$\sigma^*_{CO} \equiv (2s + 2p_z)_C - (2s - 2p_z)_O$$

$$\sigma_{CH} \equiv (2s - 2p_z)_C + \frac{1}{\sqrt{2}}(1s_1 + 1s_2)$$

$$n \equiv (2s + 2p_z)^b_O$$

$$\sigma_{CH} \equiv 2p_y + \frac{1}{\sqrt{2}}(1s_1 - 1s_2)$$

$$n_O \equiv 2p_y^{\,b}$$

$$\pi \equiv (2p_x + 2p_x)_{CO}$$

$$\pi^* \equiv (2p_x - 2p_x)_{CO}$$

Fig. 5.4. Summary of all binding and non-binding molecular orbitals of formaldehyde in the ground state. The paper represents the *xy* plane; π and π^*-orbitals are shown in side view (from Song, modified).

quantum of light needs 1.83×10^{-15} s to complete an oscillation at a given point in space. An electron moving with a speed of about 10^6 m s^{-1} traverses the distance of about 1 nm during the time of 10^{-15} s. Since the diameter of light-absorbing molecules is in the same order of magnitude, the wavelength of the exciting electromagnetic radiation corresponds fairly well with that of the electron, which can be regarded as a wave. Thus, the absorption of radiation is due to the resonance between the two waves. The movement of the electron in its orbital limits the speed of photochemical processes: below 10^{-16} s we cannot expect reactions since the electron does not move far during that time.

5.2.3 Photochemically important electronic transitions

In the following we will consider not the energy of a single electron but the total energy of a whole molecule. The levels of the excited states are drawn according to their energy above the ground state (Fig. 5.5). A molecule can be excited by various forms of energy, such as thermal or chemical energy, or by energy transfer from one molecule to the other (see Chapter 6); here we will restrict our considerations to excitation by light energy.

The second photochemical law states that one molecule can be excited by one, and only one, quantum of light. Depending on the amount of absorbed energy the molecule makes a transition from the ground state S_0 to the first (S_1) or a higher excited singlet state ($S_2, S_3, \ldots$). From the first excited singlet

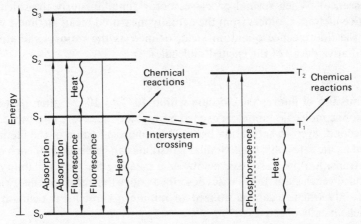

Fig. 5.5. Energy levels of the excited singlet and triplet states above the ground state. The excitation is affected by absorption of quanta with appropriate energies. The direct transition from S_0 to T_1 or T_2 is a 'forbidden' transition (highly improbable). Instead, T_1 is populated by intersystem crossing via S_1 (long dashes). The absorbed energy can be released either in the form of heat (radiationless transition, wavy lines), by fluorescence (continuous lines) or phosphorescence (short dashes) (in parts from Christen, modified).

state a molecule can be excited to a higher state by absorbing a second quantum (multiphotonic process).

However, the molecule spends only about 10^{-7} to 10^{-9} s in the first excited singlet state S_1 so that the second absorption has to occur during that time. Usually the molecule returns very fast (within 10^{-12} s) from the S_2 state to the S_1 state, whereby the energy difference is lost by internal conversion as heat. Internal conversion is an intramolecular radiationless process which does not involve a spin flip (for instance $S_1 \rightarrow S_0$ or $T_2 \rightarrow T_1$). Only in rare cases can the energy of the S_2 state be used photochemically. The molecule can return from the excited state into the ground state either by emitting the energy difference as heat or in the form of a photon, or by inducing a chemical reaction. This radiation emission is called fluorescence. The emitted fluorescence energy cannot be greater than the absorbed energy; therefore the fluorescence cannot have a shorter wavelength than the excitation wavelength. Usually it is even shifted to longer wavelengths due to energy losses; this is called Stokes' shift.

Kinetically, the fraction of the excited molecules (A*) decreases exponentially with time t:

$$\frac{A^*_t}{A^*_0} = e^{-kt}$$

From this equation we can define the radiative lifetime τ_0 within which the fractional population of the excited molecules decays to a value of $1/e$ ($\sim 37\%$).

This definition is valid only for unimolecular processes in which the relaxation of the excited molecules occurs only by one process (in this case fluorescence). When several processes occur simultaneously the calculated radiative lifetime τ_0 differs from the actually measured mean lifetime τ where ϕ_F is the fluorescence quantum yield, defined as the ratio of radiative to nonradiative decay of the excited molecules.

$$\tau = \tau_0 \, \phi_F.$$

The emission of fluorescence occurs within 10^{-9} to 10^{-6} s after excitation. Therefore a molecule cannot be excited more than once every microsecond or nanosecond, even in saturating light. The reciprocal value of the lifetime is defined as the rate constant k. Similar constants can be defined for each of the other transition processes (see below) as a measure for the availability of a molecule in each state. In the cases described above the absorbed energy is lost again; only when it can be utilized to initiate a subsequent reaction is it photochemically effective.

In addition to the relaxation to the ground state, the molecule can undergo a transition into the first excited triplet state T_1; this is called intersystem crossing. Intersystem crossing is thus another intramolecular transition which, in contrast to internal conversion, involves a spin flip. Both intersystem crossing and internal conversion are adiabatic processes. This means that they take place on the same energetic level. For example, during the $S_2 \rightarrow S_1$ transition the electron enters a higher vibronic level of S_1 (see below) on the same energy level and loses the energy difference only subsequently in the form of heat when it returns in the vibronic ground state of S_1.

The direct excitation of a molecule from its ground state S_0 into an excited triplet state is extremely rare, and is called a 'forbidden transition'. Due to the long lifetime of the triplet state, the absorbed energy can easily be utilized for a subsequent chemical reaction. The energy stored in the triplet state can be lost as heat, as in the singlet state. In this case the electron returns to its original orbital by intersystem crossing. Alternatively, a photon can be emitted; this process is called phosphorescence. In both cases the spin is reversed. Phosphorescence has a considerably longer lifetime than fluorescence and can be observed over seconds or even minutes. In addition we find a delayed fluorescence when the electron first returns to the S_1 state from which it relaxes to the ground state under fluorescence emission. When we consider all the parallel processes for relaxation, the quantum yield for one process (for example fluorescence) can be calculated by

$$\phi_F = \frac{k_F}{k_F + k_{IC} + k_{ISC}} = k_F \cdot \tau$$

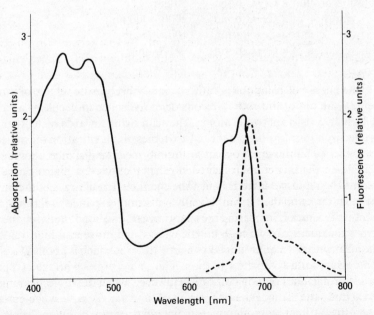

Fig. 5.6. Absorption spectrum (left ordinate, continuous line) and emission spectrum (fluorescence, right ordinate, broken line) of photosystem II from wheat chloroplasts at 77 K (from Gasanov *et al.*).

where k are the rate constants for fluorescence (F), internal conversion (IC) and intersystem crossing (ISC) and τ the measured mean lifetime.

Whether or not the incident energy is absorbed is important for a photochemical reaction. Many of the biologically important porphyrin ring systems, such as chlorophylls, cytochromes and hemoglobins, absorb both longer-wavelength quanta, which causes a transition into the S_1 state, and shorter-wavelength quanta in the Soret band, by which higher excited states S_n are generated. In the latter case the energy difference between the S_n state and the S_1 state is lost as heat, and the molecule returns to the S_1 state.

5.3   VIBRATIONAL ENERGY OF ATOMIC NUCLEI

In our previous considerations we have assumed that the nuclei of the molecule do not change their energetic state, and that only the electrons enter an excited state. The total energy of a molecule consists not only of the sum of the electronic energies but in addition there are vibrational and rotational energies of the nuclei. The following list shows the approximate wavelengths of absorbed radiation

rotation around a C—C bond	50 μm
vibration of a C—C bond	3000–5000 nm
electronic excitation	100–900 nm
dissociation of a C—C bond	$\sim$ 100 nm
ionization	< 100 nm

While the energy of rotations is rather small, vibrational energies of atomic nuclei play an important role. We consider a hydrogen molecule as a simple model which consists of two H atoms. They can swing towards each other as if connected by a spiral spring. At the two extremes of this vibration the molecule has a higher potential energy than at an intermediate distance between the nuclei. When the nuclei approach each other they need a higher energy to overcome the repulsive force between the equal electric charges of the nuclei. In the other direction they – symbolically speaking – expand the elastic spring connecting the nuclei. Increasing the energy breaks the bond. Between the two extremes the nuclei carry a high kinetic energy and move at a high velocity.

The absorption of a quantum takes about 10^{-15} s, which is about 1% of the time necessary for a complete vibration. Thus, all energies with the exception of the electronic ones remain constant. However, when the molecule remains in an excited state for an extended period of time, for example when entering the long-lived triplet state, the nuclear potential energy of the molecule can change considerably. The energy changes due to nuclear vibrations in the ground state can be shown in the form of a potential diagram (Fig. 5.7).

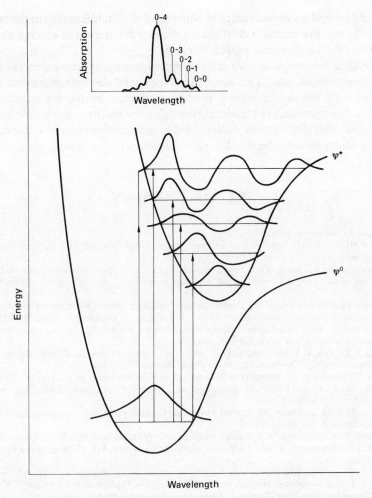

Fig. 5.7. Quantum mechanical interpretation of the Franck–Condon principle: potential diagram for two electronic states. The molecule can jump from the various energetic levels of the ground state ψ^0 to the various energetic levels of the excited state ψ^*. The transition probability depends on the time spent on each energetic level. Therefore some transitions are relatively improbable (weak absorption) while others occur frequently (strong absorption, see the schematic absorption spectrum) (from Turro, 1981, modified).

A similar potential curve can be drawn for any excited state; it is shifted with respect to the ground state since the total energy of the molecule changes when an electron enters another orbital. Within each potential curve we can define certain levels where the nuclear vibration is an oscillation with integer wave numbers (Fig. 5.7).

Speaking in terms of quantum mechanics, a nuclear vibration can be described by a wave function χ. When we call the wave function in the ground state χ_g and that in the excited state χ_e we can determine the overlap between the two by the Franck–Condon integral $\langle \chi_g | \chi_e \rangle$.

Nuclear vibrations modify the absorption spectra of substances. Instead of sharp absorption bands we usually observe broad absorption maxima which are due to the superposition of several vibronic bands. By reducing the ambient temperature (e.g. in liquid nitrogen) the number of vibronic bands is reduced. Therefore the absorption bands are much sharper and better-defined in low-temperature spectra than in room-temperature spectra.

5.4   BIBLIOGRAPHY

Textbooks and review articles

Becker, H. G. O. (ed.) *Einführung in die Photochemie*, 2. Aufl. Thieme, Stuttgart and New York (1983)

Bensasson, R. V., Land, E. J. and Truscott, T. G. *Flash Photolysis and Pulse Radiolysis: Contributions to the Chemistry of Biology and Medicine*, pp. 1–19. Pergamon Press, Oxford (1983)

Bertsch, G. F. Vibrations of the atomic nucleus. *Sci. Amer.* May, 62–73 (1983)

Bryce-Smith, D. *Photochemistry*, Vol. 15. Royal Soc. Chem., Letchworth, England (1985)

Clayton, R. K. *Photobiologie*. Bd. 1. *Physikalische Grundlagen*, Bd. 2. *Die biologischen Funktionen des Lichtes*. Verlag Chemie, Weinheim (1977)

Coyle, J. D., Hill, R. R. and Roberts, D. R. (eds) *Light, Chemical Change and Life: a Source Book in Photochemistry*. The Open University Press, Walton Hall (1982)

Dörr, F. Allgemeine Grundlagen der Photophysik und Photochemie. In: Hoppe, W., Lohmann, W., Markl, H. and Ziegler, H. (eds) *Biophysik*, 2. Aufl. Springer, Berlin, Heidelberg and New York, pp. 275–289 (1982)

Doust, T. A. M. and West, M. A. (eds) *Picosecond Chemistry and Biology*. Science Reviews Ltd, Northwood, Middlesex (1983)

Lamola, A. A. and Turro, N. J. Energy transfer and organic photochemistry. In: Leermakers, P. A. and Weissberger, A. (eds) *Techniques of Organic Chemistry*, Vol. 15. Interscience Publishers (1969)

Lim, E. C. (ed.) *Excited States*, Vols 5 and 6. Academic Press, New York and London (1982)

Rabek, J. F. *Experimental Methods in Photochemistry and Photophysics*, Part 1 and 2. J. Wiley, Chichester, New York, Brisbane, Toronto and Singapore (1982)

Rashba, E. I. and Sturge, M. D. (eds) *Excitions*. North-Holland, Amsterdam (1983)

Smith, K. C. and Hanawalt, P. C. *Molecular Photobiology*. Academic Press, New York and London (1969)

Turro, N. J. *Molecular Photochemistry*. W. A. Benjamin, Inc. Reading, Mass. (1965)

Turro, N. J. *Modern Molecular Photochemistry*. Benjamin/Cummings, Menlo Park, Calif. (1978); student edition (1981)

Further reading

Andreyeva, N. E. and Chibisov, A. Role of triplet exciplex in the reaction of electron transfer. *Biofizika* **25**, 965–971 (1980)

Heelis, P. F., De la Rosa, M. A. and Phillips, G. O. A laser flash photolysis study of the photoreduction of the lumiflavin triplet state. *Photobiochem. Photobiophys.* **9**, 57–63 (1985)

Innes, K. K. Vibronic coupling as a major cause of the anharmonicity of antisymmetric stretching in the ground state of NO_2. *J. Mol. Spectr.* **96**, 331–335 (1982)

Krasnovsky, A. A. Jr. Delayed fluorescence and phosphorescence of plant pigments. *Photochem. Photobiol.* **36**, 733–741 (1982)

Merrifield, R. E. and Simmons, H. E. Topology of bonding in electron systems. *Proc. Natl. Acad. Sci.* **82**, 1–3 (1985)

Nagata, C. and Aida, M. Ab initio molecular orbital study of the interaction of Li^+, Na^+ and K^+ with the pore components of ion channels: Consideration of the size, structure and selectivity of the pore of the channels. *J. Theor. Biol.* **110**, 569–585 (1984)

Nassau, K. The causes of color. *Sci. Am.* **243**, 106–123 (1980)

6 Photochemical reactions and energy transfer

By electronic excitation a molecule enters an excited state with an energy and charge distribution of the electrons different from that in the ground state. This transition changes its molecular properties such as bond strength, the pK of certain atoms or groups, the redox potential and the electron affinity. These changes can induce a number of biologically relevant reactions which pertain to either one molecule (unimolecular processes) or several molecules (bimolecular, multimolecular processes).

6.1 PHOTODISSOCIATION

In the simplest case a molecule can be dissociated after the absorption of a quantum. The excitation energy is used to break the bond. This process occurs, for example, during the photodissociation of silver halides by which silver is produced during irradiation of a photographic film or paper. Both radical dissociations (homolytic), such as the releasing of CO from ketones, and – especially in polar solvents – ionic dissociations, such as the release of CH^- from cyanides, have been described. Formally a photodissociation can be described by the equation:

$$A + h\nu \rightarrow A^* \rightarrow \rightarrow \rightarrow A_1 + A_2.$$

6.2 ISOMERIZATION AND CONFORMATIONAL CHANGE

Electronic excitation can result in the photoisomerization of a molecule. One of the best-known examples is rhodopsin. In the dark, the chromophoric group is an 11-*cis* isomer. Upon absorption of a quantum the molecule changes to the all-*trans* form (Fig. 6.1). During isomerization we find reversible changes in the absorption which are responsible for a photobleaching or a light-induced absorbance change of the molecule. This photochromic effect is well known in nitro compounds and in aniline dyes.

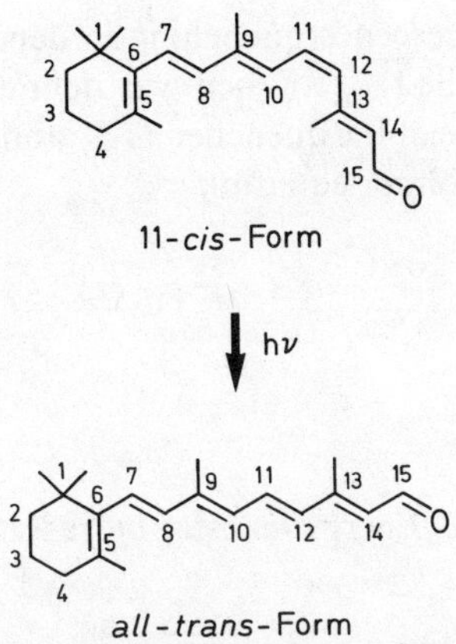

11-*cis*-Form

↓ hν

all-*trans*-Form

Fig. 6.1. Photoisomerization of retinal, the chromophoric group of rhodopsin. By absorption of a quantum the 11-*cis* form changes to the all-*trans* form (from Rosenfeld *et al.*, 1977, modified).

In addition to isomerizations and tautomeric changes, we find light-induced formation or splitting of rings, enolization or a rearrangement of substituents. The photobiologist is especially interested in changes of the chromophoric group which induce secondary conformational changes of the protein it is linked to. By this process, for example, pores or channels can be opened in a membrane which allows an ionic gradient to equilibrate (see below). Conformational changes can also be due to a light-induced polarization or a change in the dipole moment of a photoreceptor molecule which can alter the conformation of the surrounding membrane proteins.

6.3 ENERGY TRANSFER BETWEEN MOLECULES

The molecule which has entered an excited state upon absorption of a quantum is not necessarily responsible for the following photochemical processes. Instead, it can transfer its energy to other molecules.

6.3.1 Fluorescence quenching

The quantum yield of fluorescence ϕ_F is lowered when bimolecular processes such as quenching occur simultaneously with fluorescence and radiationless processes (intersystem crossing and internal conversion, see Section 5.2.3).

$$\phi_F = \frac{k_F}{k_F + k_{IC} + k_{ISC} + k_Q[Q]} = k_F \cdot \tau$$

The rate constant of fluorescence quenching in dependence of the quencher concentration $[Q]$ is called k_Q. When we define the quantum yield of fluorescence in the absence of the quencher as ϕ_F^o and the measured lifetime as τ_M we obtain the Stern-Volmer equation

$$\frac{\phi_F^o}{\phi_F} = 1 + \tau_M \cdot k_Q[Q]$$

6.3.2   Energy transfer by radiation

Formally, we can write the energy transfer from an excited donor molecule D to an acceptor A as

$$D^* + A \rightarrow D + A^*.$$

During this energy transfer the donor emits a quantum which is absorbed by the acceptor

$$D^* + A \rightarrow D + h\nu + A \rightarrow D + A^*.$$

One of the prerequisities is, of course, that the fluorescence emission spectrum of the donor molecule overlaps with the absorption spectrum of the acceptor molecule. When we integrate over the overlap between the spectra and normalize to one we calculate the transfer probability as a number between zero and one. This calculation already includes the fluorescence quantum yield, the concentrations of the donor and the acceptor as well as the absorption coefficient of the acceptor. Due to energetic reasons the radiation is shifted toward longer wavelengths during each step. As well as fluorescence, delayed fluorescence and phosphorescence can also be reabsorbed by other molecules.

6.3.3   Radiationless energy transfer

6.3.3.1   ELECTRON EXCHANGE

Bimolecular chemical interactions often occur during collisions of electron clouds of two partners. During the spatial overlap of the clouds, electrons can be exchanged. The exchange of the donor and acceptor electron can occur simultaneously or in sequence by forming ionic radicals (charge transfer).

6.3.3.2   EXCIMERS AND EXCIPLEXES

When the two partners – the donor molecule in the excited state and the acceptor molecule in the ground state – are sufficiently polarized they can form a metastable complex in the excited state. This complex is called an excimer when the two partners are identical, and it is called an exciplex when they are different. The metastable complex has properties different from the excited donor molecule: the absorption maximum of the complex differs from that of the individual molecules, so that the existence of photochemically active dimeric or polymeric complexes can be seen in the emission spectrum as shown in the example of the pyrene fluorescence in n-heptane (Fig. 6.2). At low concentrations (10^{-5} M) too few excited molecules collide with molecules in the ground state to produce a noticeable number of excimers. When we increase the concentration 100-fold the probability for collisions between molecules is far greater, so that the incident quanta are reemitted almost exclusively by excimers.

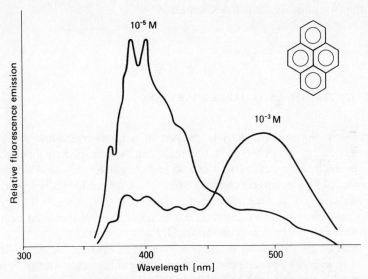

Fig. 6.2. Fluorescence spectra of pyrene in heptane at concentrations of 10^{-3} M and 10^{-5} M. In the higher concentration excimers are formed, the fluorescence of which is shifted to longer wavelengths than that of the individual molecules (from Turro, 1978).

6.3.3.3   ENERGY TRANSFER BY RESONANCE

Energy can be exchanged between molecules even over considerable distances. Förster regarded the excited molecule as a dipole oscillator. Instead of emitting the excitation energy in the form of an electromagnetic wave it can be directly transferred to a neighboring molecule by excitation resonance transfer. The transfer probability is inversely proportional to the sixth power of the distance between the molecular centers. The energy transfer by resonance can be visualized with the behavior of linked pendulums. The transfer is only possible when the frequencies of the donor and the acceptor are in resonance. Thus, also in this case the efficiency of the intermolecular energy transfer depends on the overlap between the donor and acceptor spectra (see Section 6.2.1).

Transfer between equal molecules occurs without any energy losses. When the excited state of the acceptor is energetically lower than that of the donor the energy difference is lost as heat. An 'uphill' energy transfer is only possible in exceptional cases when the energy difference can be compensated by vibrational or rotational energies of the acceptor molecule. Resonance transfer is discussed as a possible mechanism for the energy transfer from the photosynthetic accessory pigments to the reaction centers of the two photosystems (see Chapter 9).

6.3.3.4   EXCITATION AND CHARGE MIGRATION

When the molecules are densely packed in a quasi-crystalline array the excitation energy absorbed by a molecule can be transported through the lattice by exciton migration. In a crystal each negatively charged electron is compensated by a positively charged 'hole' which we can formally regard as a positively charged particle. Both can migrate together through the crystal. When the electron is in an excited state the exciton (hole plus excited electron) travels through the lattice as an electrically neutral particle.

In contrast to the joined migration, either the electron or the hole can recombine with the opposite charge of an adjacent molecule. In this case we are left with a positive or negative charge which carries the excitation energy through the molecular array.

One of the prerequisites for both exciton and charge transfer is a low ionization potential so that the excited electron can be separated from the molecule. Figure 6.3 shows the absorption spectra during charge transfer from a donor D (various enol ethers) to an acceptor A (tetracyanoethylene). The transfer can formally be written as

$$D + A + h\nu \rightarrow D^* + A \rightarrow D^+ + A^-.$$

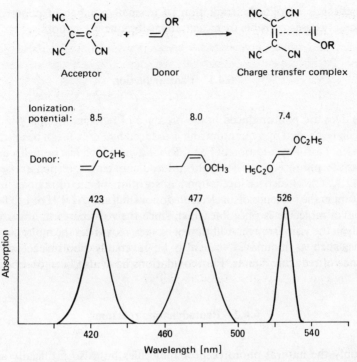

Fig. 6.3. Absorption spectra indicating a charge migration from a donor (various enol ethers) to an acceptor (tetracyanoethylene). The wavelengths of the absorption maxima correspond with the ionization potentials (from Turro, 1978, modified).

6.4 BIMOLECULAR PHOTOCHEMICAL REACTIONS

Similar to unimolecular photodissociation, excitation energy can be utilized to split a molecule, although in this case the substrate molecule S is not identical to the excited donor molecule D*

$$D + h\nu \rightarrow D^*$$
$$D^* + S \rightarrow D + I^+ + I^-.$$

The resulting substrate ions I can react and form a stable end-product. The reaction can also result in the formation of radicals R instead of ions

$$D^* + S \rightarrow D + R_1^{\cdot} + R_2^{\cdot}.$$

The formed radicals can catalytically induce chain reactions so that a large number of stable end-products are produced per excited molecule. Thus the light effect is amplified: irradiation of a small number of photoreceptor molecules causes a visible or measurable subsequent reaction.

6.4.1   Photoreduction

Some dyes are photoreduced in the presence of reducing agents. Methylene blue can be excited by a quantum and is reduced by Fe^{2+}. It can be returned to its original form by addition of Fe^{3+}. Riboflavin, which has been discussed as a possible photoreceptor for light-induced movement responses (see Section 12.1.1.2), is also reduced upon absorption of an ultraviolet or blue quantum in the presence of an electron donor such as NADPH or EDTA. The riboflavin molecule is photobleached, since the reduced form absorbs only weakly in the visible range. Addition of oxygen reoxidizes the molecule. Other flavins, such as lumiflavin, can also be reversibly photobleached in the presence of reducing agents. Photooxidations have also been observed.

6.4.2   Photodynamic reactions

Not only the natural photoreceptor molecules but also artificially applied substances can induce light-dependent reactions in living organisms. These (mostly destructive) reactions induced by irradiation of endogenous and exogenous sensitizer substances are called photodynamic actions. Even a small concentration of eosin kills the ciliate *Paramecium* in light, while the organism survives in darkness. Photodynamic actions are also found in higher animals: when cows feed on St John's wort (*Hypericum*) or buckwheat (*Fagopyrum*) species, the hypericine or fagopyine contained in the plants induces severe photodynamic damages, which can even be fatal when the animals are exposed to light for an extended duration. Chlorophylls and related compounds can also act as effective photosensitizers. In 1913 the physician Myer-Betz injected himself with 200 mg of hematoporphyrin. When exposed to light he suffered from a severe erythema (see Chapter 14) and the light sensitivity lasted for several months.

Most photodynamic effects start from the first excited triplet state. The excited photosensitizer P can detach a hydrogen atom or electron from a reducing substrate S and form a half-reduced free radical. Two radicals react to one oxidized and one reduced molecule. The latter reduces molecular oxygen to hydrogen peroxide.

$$2 \, PH^{\cdot} \rightarrow P + PH_2$$
$$PH_2 + O_2 \rightarrow P + H_2O_2$$

A second possible reaction has been described in which the excitation energy is transferred to oxygen. In contrast to most other molecules, oxygen is in the triplet state when in the ground state. By reaction with an excited photosensitizer singlet oxygen is formed

$$^3P^* + {}^3O_2 \rightarrow P + {}^1O_2^*.$$

Singlet oxygen can readily oxidize many biological substances such as amino acids and unsaturated fatty acids.

6.5 SECONDARY REACTIONS

After a quantum has been absorbed by an appropriate photoreceptor, and the excitation energy has possibly been transported via several intermediate molecules to a terminal acceptor, the energy is converted into another form and used for chemical work. The excited states have only a limited lifetime of 10^{-13} to 10^{-9} a in the excited singlet state and 10^{-4} to 1 s in the excited triplet state. Therefore the energy difference caused by absorption of radiation needs to be used within a relatively short time. Often the reaction partner is adjacent to the absorbing pigment. In biological systems photoreceptor molecules are often linked as chromophoric groups to membrane-bound proteins. Interactions with neighboring molecules cause changes in the absorption characteristics (position and height of maxima) as compared with pure solutions. The location of photoreceptor molecules in membranes guarantees an effective transport and utilization of the excitation energy and additionally allows the separation of charges and reaction products in different compartments (see Section 6.3.4). Often the photoreceptor molecule undergoes a reaction cycle involving one or more intermediates. At one point in the cycle the molecule is excited, and at another it relaxes and initiates a subsequent reaction.

6.5.1 Proton gradient and membrane potential

The pK_a of a functional group of a photoreceptor can change drastically after light absorption. This changes the affinity of this group toward protons and electrons. When the protein component undergoes a conformational change, protons could be accepted preferentially from one side of the membrane and subsequently released at the other (Fig. 6.4).

In his chemiosmotic hypothesis Mitchell has proposed a mechanism by which the energy stored in a proton gradient can be converted into chemical energy in the form of ATP. In photosynthesis, however, the proton

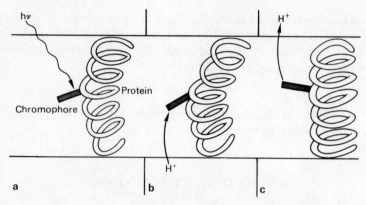

Fig. 6.4. Vectorial proton transport through a biomembrane. After absorption of a quantum the chromophoric group changes its pK_a (a), which leads to a proton uptake (b). After a subsequent conformational change of the protein the proton is released at the opposite side of the membrane (c).

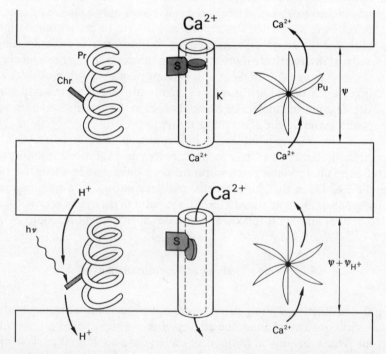

Fig. 6.5. Schematic presentation of the ion transport through a membrane. Energy-consuming pumps (Pu) transport ions (such as Ca^{2+}) through the membrane to the outside, which generates an ion gradient and an electrical potential Ψ. When a membrane-bound photoreceptor (protein Pr with chromophoric group Chr) causes a proton transport in light, the resulting proton gradient Ψ_{H^+} changes the resting potential across the membrane. This potential change is detected by a sensor S of a voltage-dependent ion channel K and causes an opening of the channel. The sudden influx of Ca^{2+} ions along the previously established gradient causes a massive potential change measured across the membrane.

transport is not carried out by the photoreceptor molecule itself but indirectly by an electron transport chain (see Chapter 9). The resulting proton gradient is used to power an ATP synthase, an enzyme which phosphorylates ADP to ATP.

Since protons are charged ions vectorial proton transport not only results in a pH difference across the membrane but also generates an electrical potential $\Delta\psi$. A potential gradient can be measured across each membrane due to ionic concentration differences between the two compartments separated by the membrane. Active ion pumps generate a resting potential specifically in the plasma membrane (the outer membrane of a cell). The internal and external concentrations of an ion can differ as much as 1000-fold. In addition to ion pumps we find passive ion channels or gates in the membrane, which are usually closed. These channels can be visualized as tubular proteins (Fig. 6.5).

When, by the action of a membrane-bound photoreceptor, a proton gradient is generated, the increase in the potential $\Delta\psi$ can induce a conformational change of ion channels, which open and allow a passive ion current through the membrane along the previously established ion gradient. The massive influx of ions changes the membrane potential even more than the original proton gradient, which results in a considerable amplification of the photoeffect.

6.6 BIBLIOGRAPHY

Textbooks and review articles
Coyle, J. D., Hill, R. R. and Roberts, D. R. (eds) *Light, Chemical Change and Life: a Source Book in Photochemistry*. The Open University Press (1982)
Daynes, R. A. and Spikes, J. A. (eds) *Experimental and Clinical Photoimmunology*, Vol. 1. *Photobiology and Basic Immunology*. (1983)
Dörr, F. Allgemeine Grundlagen der Photophysik und Photochemie. In: Hoppe, W., Lohmann, W., Markl, H. and Ziegler, H. (eds) *Biophysik*, 2nd edn. Springer, Berlin, Heidelberg and New York (1982)
Förster, T. *Fluoreszenz organischer Verbindungen*. Vandenhoeck and Ruprecht, Göttingen (1951)
Kan, R. O. *Organic Photochemistry. Series in Advanced Chemistry*, McGraw-Hill, New York (1966)
Kuhn, H. Energieübertragungsmechanismen. In: Hoppe, W., Lohmann, W., Markl, H. and Ziegler, H. (eds) *Biophysik*. 2nd edn. Springer, Berlin, Heidelberg and New York (1982)
Lakowicz, J. R. *Principles of Fluorescence Spectroscopy*. Plenum Press, New York and London (1983)
Lamola, A. A. and Turro, N. J. Energy transfer and organic photochemistry. In: Leermakers, P. A. and Weissberger, D. (eds) *Techniques of Organic Chemistry*, Vol. 15. Interscience Publishers (1969)
Mitchell, P. Chemiosmosis: term of abuse. *Trends Biochem.* **9**, 205 (1984)
Mitchell, P. Molecular mechanics of protonmotive F_oF_1 ATPases. Rolling well and turnstile hypothesis. *FEBS Lett.* **182**, 1–7 (1985)
Mitchell, P., Mitchell, R., Moody, A. J., West, I. C., Baum, H. and Wrigglesworth, J. M. Chemiosmotic coupling in cytochrome oxidase. *FEBS Lett.* **188**, 1–7 (1985)
Rashba, E. I. and Sturge, M. D. (eds) *Excitons*. Elsevier, North-Holland, Amsterdam (1983)
Smith, K. C. *The Science of Photobiology*. Plenum Press, New York and London (1977)

Song, P.-S. The electronic spectroscopy of photoreceptors (other than rhodopsin). In: Smith, K. C. (ed.) *Photochemical and Photobiological Reviews*, Vol. 7. Plenum Press, New York and London (1983)

Song, P.-S. Primary molecular events in aneural cell photoreceptors. In: Colombetti, G., Lenci, F. and Song, P.-S. (eds) *Sensory Perception and Transduction in Aneural Organisms*. Plenum Press, New York and London, pp. 47–59 (1985)

Tsien, R. W. Calcium channels in excitable cell membranes. *Ann. Rev. Physiol.* 45, 41–58 (1983)

Turro, N. J. *Modern Molecular Photochemistry*. Benjamin/Cummings Menlo Park, Calif. (1978)

Further reading

Andreyeva, N. Y. and Chibisov, A. K. Role of the triplet exciplex in the reaction of electron transfer. *Biophysics* 25, 987–994 (1980)

Bensasson, R. V., Land, E. J., Liu, R. S. H., Lo, K. K. N. and Truscott, T. G. Triplet states of isomers of the C_{15}-aldehyde and the C_{18}-ketone: lower homologues of retinal. *Photochem. Photobiol.* 39, 263–265 (1984)

Butler, W. L. Exciton transfer out of open photosystem II reaction centers. *Photochem. Photobiol.* 40, 513–518 (1984)

Chauvet, J.-P., Bazin, M. and Santus, R. On the triplet–triplet energy transfer from chlorophyll to carotene in triton X-100 micelles. *Photochem. Photobiol.* 41, 83–90 (1985)

Clapp, R. E. Zipper transition in an alpha-helix: a mechanism for gating of voltage-sensitive ion channels in a biological membrane. *J. Theor. Biol.* 104, 137–158 (1983)

Creed, D. and Caldwell, R. A. Photochemical electron transfer reactions and exciplexes. *Photochem. Photobiol.* 41, 715–739 (1985)

Cunningham, M. L., Johnson, J. S., Giovanazzi, S. M. and Peak, M. J. Photosensitized production of superoxide anions by monochromatic (290–405 nm) ultraviolet irradiation of NADH and NADPH coenzymes. *Photochem. Photobiol.* 42, 125–128 (1985)

Kireyev, V. B., Skachkov, M. P., Trukhan, E. M. and Filimonov, D. A. Charge transfer in the primary processes of photosynthesis. *Biophysics* 26, 8–12 (1981)

Knox, J. P. and Dodge, A. D. Photodynamic damage to plant leaf tissue by rose bengal. *Plant Sci. Lett.* 37, 3–7 (1984)

Knox, J. P. and Dodge, A. D. Isolation and activity of the photodynamic pigment hypericin. *Plant Cell Environ.* 8, 19–25 (1985)

Kolubsyev, T., Geacintov, N. E., Paillotin, G. and Breton, J. Domain sizes in chloroplasts and chlorophyll–protein complexes probed by fluorescence yield quenching induced by singlet-triplet exciton annihilation. *Biochim. Biophys. Acta* 808, 66–76 (1985)

Krasnovsky, A. A. Jr. Delayed fluorescence and phosphorescence of plant pigments. *Photochem. Photobiol.* 36, 733–741 (1982)

Kudzmauskas, S. and Valkunas, L. A theory of excitation transfer in photosynthetic units. *J. Theor. Biol.* 105, 13–23 (1983)

Mauzerall, D., Ho, P. P. and Alfano, R. R. The use of short lived fluorescent dyes to correct for artifacts in the measurements of fluorescence lifetimes. *Photochem. Photobiol.* 42, 183–186 (1985)

Mitchell, P. Protonmotive chemiosmotic mechanisms in oxidative and photosynthetic phosphorylation. *TIBS* 3, N58–N61 (1978)

Mohanty, P., Hoshina, S. and Fork, D. C. Energy transfer from phycobilins to chlorophyll *a* in heat-stressed cells of *Anacystis nidulans:* characterization of the low temperature 683 nm fluorescence emission band. *Photochem. Photobiol.* 41, 589–596 (1985)

Nultsch, W. and Kumar, H. D. Effects of quenching agents on the photodynamically-induced chemotactic response of the colorless flagellate *Polytomella magna. Photochem. Photobiol.* 40, 539–543 (1984)

Rosenfeld, T., Honig, B., Ottolenghi, M., Hurley, J. and Ebrey, T. G. *Cis–trans* isomerization in the photochemistry of vision. *Pure Appl. Chem.* 49, 341–351 (1977)

Salet, C., Bazin, M., Moreno, G. and Favre, A. 4-Thiouridine as a photodynamic agent. *Photochem. Photobiol.* 41, 617–619 (1985)

Schmidt, W. Inhibition of bluelight-induced, flavin-mediated membraneous redoxtransfer by xenon. *Z. Naturf.* 40c, 451–453 (1985)

Setty, O. H., Hendler, R. W. and Shrager, R. I. Simultaneous measurements of proton motive force pH, membrane potential, and H^+/O ratios in intact *Escherichia coli*. *Biophys. J.* **43**, 371–381 (1983)

Wehrmeyer, W., Wendler, J. and Holzwarth, A. R. Biochemical and functional characterization of a peripheral phycobilisome unit from *Porphyridium cruentum*. Measurement of picosecond energy transfer kinetics. *Europ. J. Cell Biol.* **36**, 17–23 (1985)

7 Bioluminescence and chemiluminescence

Reports of marine photoluminescence date back to Anaximenes (about A.D. 500); Plinius describes the bioluminescence of the shell *Pholas* and of medusae. The ability to emit light can be deduced from the names of many organisms: *Photobacterium phosphoreum* (bacteria), *Pyrocystis* and *Noctiluca* (peridineae), *Pelagia noctiluca* (medusae), *Pyrocypris* and *Metridia lucens* (crustaceans), *Watasenia scintillans* and *Symplectotheutis luminosa* (molluscs), *Pyrosoma* (tunicate), *Photoblepharon* (fish) and *Photuris pennsylvanica* and *Pyrophorus* (insects; Fig. 7.1). This list demonstrates the wide taxonomic

Fig. 7.1. Firefly (*Pyrophorus*), from the American tropics, with two light organs on its thorax behind the eyes. In addition, the firefly has another light organ on the abdomen (original photograph courtesy of D. Haarhaus).

distribution of bioluminescence, which has been found in bacteria, fungi, unicellular algae and in most of the animal phyla, some of which include hundreds of luminescent species. Some higher organisms such as fish do not generate bioluminescence by themselves but rather use symbiotic bacteria (*Photobacterium*) in their luminous organs.

7.1   CHEMILUMINESCENCE

The phenomena described in this chapter differ from the other photochemical and photobiological processes in that they do not occur as a consequence of the absorption of quanta. Instead, molecules are excited exclusively by chemical reactions. The following example illustrates the process. Cyclic peroxides such as 1,2-dioxetane break into two carbonyl fragments even at relatively low temperatures ($< 150°C$). One of the products will be found to be in an excited state with a high probability. In dioxetanes the reaction product is usually excited to the triplet rather than the singlet state. The ratio of singlet to triplet excitation depends on the electron configuration of the product, in other words on whether n,π^* or π,π^* transitions dominate. The energy of the excited state is released during relaxation of the molecule into the ground state in the form of fluorescence. More than 40 stable dioxetanes have already been synthesized which show fluorescence. The underlying chemical reactions differ slightly but in each case the degradation of a cyclic peroxide plays an important role.

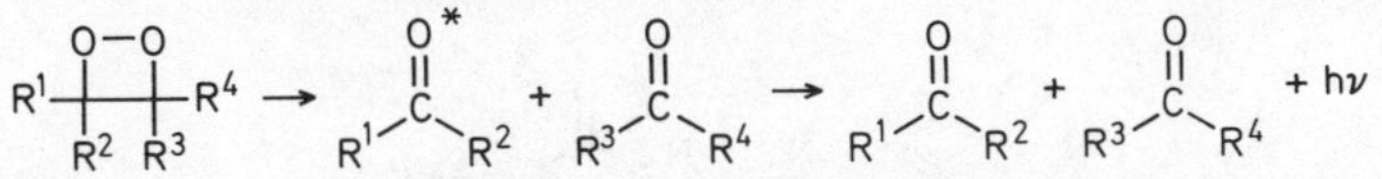

Fig. 7.2. Splitting of dioxetanes into two carbonyl fragments, one of which is in an excited state from which it returns into the ground state, emitting light (from Hastings and Wilson, modified).

Recently the measurement of very weak chemiluminescence became possible by using highly sensitive photomultipliers which can detect single photons. This technique allows one to measure the extremely weak biological luminescence emitted by, for example, liver microsomes or leukocytes. This is often used for a number of biochemical and clinical tests. In some cases a bioluminescence immunoassay can replace the widely used, but more complicated, radio immunoassay. Aequorin is a luminescence indicator for calcium fluxes in physiological systems. (Aequorin is named after the medusa *Aequorea* from which the luciferin has been isolated.) The luminescence is a linear function of the calcium concentration over a wide range, and therefore this indicator allows the experimenter to quantify the calcium level of a system after calibration of the photomultiplier current.

7.2   BIOLUMINESCENCE

Starting from the definition of chemiluminescence we can describe bioluminescence as an enzymatically catalyzed chemiluminescence in biological systems.

The energy generated by the reaction has to correspond to the shortest emitted wavelength. In *Renilla* this value is 356 kJ/mol since the shortest wavelength is 340 nm in the UV-A band. In all investigated organisms a luciferin is the substrate for a luciferase which may have evolved from a more general oxygenase. It is interesting to note that different taxonomic groups have developed chemically different luciferins (Fig. 7.3). However, the same luciferin is found in taxonomically unrelated groups, such as the luciferin of coelenterates which is also found in crustaceans, molluscs and fishes. An alternative explanation is that the luciferin is taken up from a common food supply.

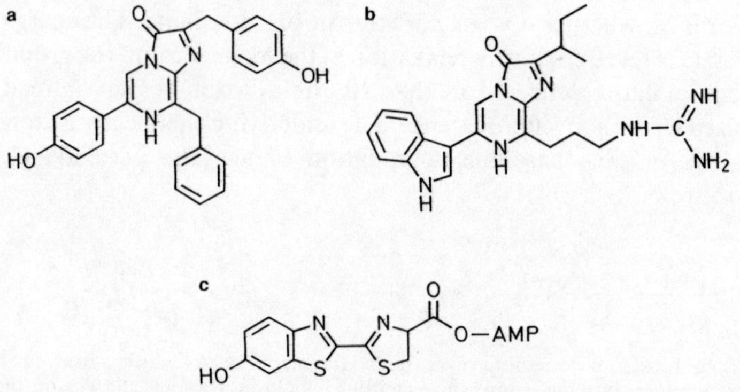

Fig. 7.3. Luciferins of (a) Coelenterates (*Renilla*, *Aequoria*); (b) *Cypridia*; and (c) fireflies (from Hastings and Wilson, and McCapra and Hart, 1980, modified).

7.2.1   Bacteria

According to the different luciferins the biochemical reactions differ. In bacteria, flavin mononucleotide (FMN) is excited (Fig. 7.4). In the presence of oxygen the reduced FMN is bound to a luciferase and enters an excited state via several intermediates. FMN is released in the oxidized form emitting a photon. One oxygen atom reacts with protons to form water. The second oxygen atom is bound by the long-chain aliphatic aldehyde which is oxidized to the corresponding carboxylic acid. *In vivo* both FMN and the produced carboxylic acid are reduced by $NAD(P)H + H^+$ (Fig. 7.5).

Even though the luciferases of different bacteria vary, there are a number of similarities. The enzymes of all investigated species consist of heterodimers with two different subunits with molecular weights of 41,000 (α) and 38,000 (β), respectively. The active center is located in the α-subunit.

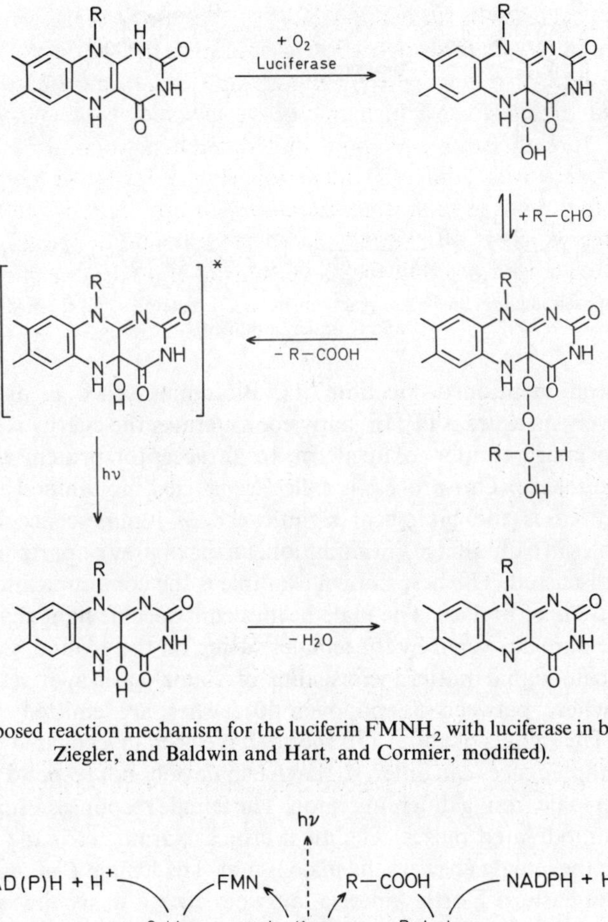

Fig. 7.4. Proposed reaction mechanism for the luciferin $FMNH_2$ with luciferase in bacteria (from Ziegler, and Baldwin and Hart, and Cormier, modified).

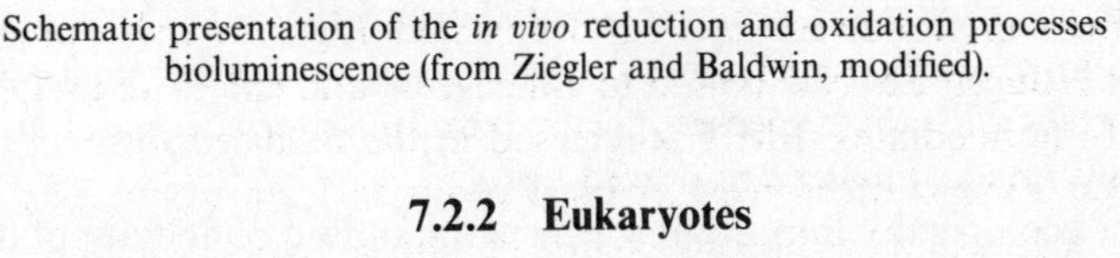

Fig. 7.5. Schematic presentation of the *in vivo* reduction and oxidation processes in bacterial bioluminescence (from Ziegler and Baldwin, modified).

7.2.2 Eukaryotes

In eukaryotes also relatively large organic molecules serve as luciferins. The luciferase catalyzes both an oxidation and a decarboxylation (Fig. 7.6). While the anthozoa have a 'classical' luciferin–luciferase system, the luminescent hydrozoa and scyphozoa possess an oxygen-independent and calcium-activated monomeric photoprotein. The biochemical use of aequorin has

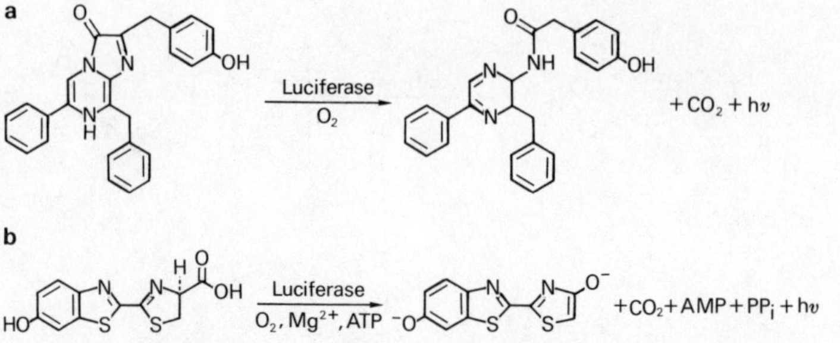

Fig. 7.6. Proposed luciferin–luciferase reactions in (a) Coelenterates and (b) fireflies (from Hart and Cormier, modified).

already been mentioned (Section 7.1). Bioluminescence is also used to quantitatively measure ATP. In many coelenterates the energy is transferred from the primary emitter, oxiluciferin, to an acceptor protein which emits green fluorescence. This process is called 'sensitized bioluminescence'.

In many cases the biological significance of luminescence is obscure. Possible roles in camouflage, intimidation, luring of prey or partner attraction are being discussed. The best-known example is the communication between sexual partners of fireflies. The male beetles emit specific light signals during flight, which are answered by the females sitting on the ground. Each species has a specific signal pattern consisting of single flashes or a modulated emission where between 5 and over 40 flashes are emitted per second (Fig. 7.7). The signals of about 130 species have been investigated worldwide. Even geographic races can differ, so that a female does not respond to the flash pattern of a male from a different region. The female responses usually consist of short unmodulated pulses. The most critical parameter is the delay time after which the female answers the male signal. The female *Photinus ignitus*, a species from eastern North America, answers after a delay time of 3 s. The delay time (as well as the flash frequency of the males) depends on the temperature.

While most adult fireflies are not carnivorous, some species of the genus *Photuris* have developed a remarkable predation behavior. The females have developed an extensive repertoire of mimicry and answer the male signals of several species with their species-specific pattern. After successful attraction the (usually smaller) males are preyed upon.

Another ecologically important factor is the timing of activity of different species. Most fireflies fly shortly after sunset during a period of about 15 to 20 min, which is observed very accurately. Each species has its own time niche. The period of activity of *Photinus tanytoxus* commences about 5 min after the related *P. collustrans* has stopped flying. This precise timing increases the chance for sexual partners to find each other, while the short time of activity reduces the risk of being eaten by predators.

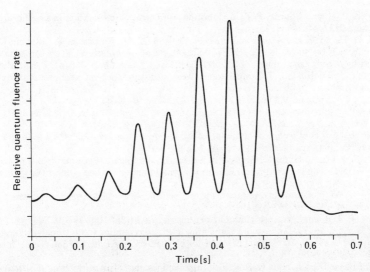

Fig. 7.7. Light signals of a male *Photinus evanescens* from Jamaica. At 21.7°C up to 11 pulses are emitted with a frequency of 20 Hz (from Lloyd, modified).

7.3 BIBLIOGRAPHY

Textbooks and review articles

Baldwin, T. O., Johnston, T. C. and Swanson, R. Recent progress in bioluminescence: cloning of the structural genes encoding bacterial luciferase, analysis of the encoded sequences, and crystallization of the enzyme. In: Bray, R. C., Engel, P. C. and Mayhew, S. G. (eds) *Flavins and Flavoproteins*. Walter de Gruyter, Berlin and New York, pp. 345–358 (1984)

Buskey, E. J. and Swift, E. Behavioral responses of oceanic zooplankton to simulated bioluminescence. *Biol. Bull.* **168**, 263–275 (1985)

Cadenas, E. Biological chemiluminescence. *Photochem. Photobiol.* **40**, 823–830 (1984)

Lloyd, J. E. Mimicry in the sexual signals of fireflies. *Sci. Amer.* July, 111–117 (1981)

Mielenz, K. D. (ed.) *Measurement of Photoluminescence*. Academic Press, New York (1982)

Shimomura, O. Bioluminescence. *Photochem. Photobiol.* **38**, 773–779 (1983)

Slawinska, D. and Slainski, J. Biological chemiluminescence. *Photochem. Photobiol.* **37**, 709–715 (1983)

Widder, E. A., Latz, M. I. and Case, J. F. Marine bioluminescence spectra measured with an optical multichannel detection system. *Biol. Bull.* **165**, 791–810 (1983)

Ziegler, M. M. and Baldwin, T. O. Biochemistry of bacterial bioluminescence. *Curr. Top. Bioenergetics* **12**, 65–113 (1981)

Further reading

Ahmad, R., Wu, Z. and Armstrong, D. A. Reactions of lumiflavin and lumiflavin radicals with ·CO₂⁻ and alcohol radicals. *Biochem.* **22**, 1806–1810 (1983)

Bogan, D. J., Durant, J. L. Jr, Sheinson, R. S. and Williams, F. W. Formation and chemiluminescent decomposition of dioxetanes in the gas phase. *Photochem. Photobiol.* **30**, 3–15 (1979)

Campa, A. and Cilento, G. Low-level luminescence from microsomes exposed to enzymatic systems that generate triplet species. *Arch. Biochem. Biophys.* **235**, 673–678 (1984)

Cobbold, P. H. and Bourne, P. K. Aequorin measurement of free calcium in single heart cells. *Nature* **312**, 444–446 (1985)

Cohn, D. H., Mileham, A. J., Simon, M. I., Nealson, K. H., Rausch, S. K., Bonam, D. and Baldwin, T. O. Nucleotide sequence of the luxA gene of *Vibrio harveyi* and the complete amino acid sequence of the subunit of bacterial luciferase. *J. Biol. Chem.* **260**, 6139–6146 (1985)

Dahlgren, C. and Briheim, G. Comparison between the luminol-dependent chemiluminescence of polymorphonuclear leukocytes and of the myeloperoxidase–HOOH system: Influence of pH, cations and protein. *Photochem. Photobiol.* **41**, 605–610 (1985)

Dunlap, P. V. Physiological and morphological state of the symbiotic bacteria from light organs of ponyfish. *Biol. Bull.* **167**, 410–425 (1984)

Dunlap, P. V. Osmotic control of luminescence and growth in *Photobacterium leiognathi* from ponyfish light organs. *Arch. Microbiol.* **141**, 44–50 (1985)

Girotti, S., Roda, A., Ghini, S., Grigolo, B., Carrea, G. and Bovara, R. Continuous flow analyses of NADH using bacterial bioluminescent enzymes immobilized on nylon. *Analyt. Lett.* **17**, 1–12 (1984)

Grogan, D. W. Interaction of respiration and luminescence in a common marine bacterium. *Arch. Microbiol.* **137**, 159–162 (1984)

Herring, P. J. Bioluminescence of marine organisms. *Nature* **267**, 788–793 (1977)

Inaba, H., Shimizu, Y., Tsuji, Y. and Yamagishi, A. Photon counting spectral analyzing system of extra-weak chemi- and bioluminescence for biochemical application. *Photochem. Photobiol.* **30**, 169–175 (1979)

Ismail, S. A. and Santhanam, K. S. V. Electrobioluminescence of an earthworm. *Bioelectrochem. Bioenerg.* **12**, 535–540 (1984)

Johnson, C. H., Inoue, S., Flint, A. and Hastings, J. W. Compartmentalization of algal bioluminescence: autofluorescence of bioluminescent particles in the dinoflagellate *Gonyaulax* as studied with image-intensified video microscopy and flow cytometry. *J. Cell Biol.* **100**, 1435–1446 (1985)

Johnson, P. C., Ware, J. A., Cliveden, P. B., Smith, M., Dvorak, A. M. and Salzman, E. W. Measurement of ionized calcium in blood platelets with the photoprotein aequorin. *J. Biol. Chem.* **260**, 2069–2076 (1985)

Kurfürst, M., Ghisla, S. and Hastings, J. W. Bioluminescence emission from the reaction of luciferase-flavin mononucleotide radical with O_2^{-}. *Biochem.* **22**, 1521–1525 (1983)

Kuwae, T. and Kurata, M. Chemical and biological properties of lipopolysaccharides from symbiotic luminous bacteria from several luminous marine animals. *Microbiol. Immunol.* **27**, 137–149 (1983)

Lee, J. Sensitization by lumazine proteins of the bioluminescence emission from the reaction of bacterial luciferases. *Photochem. Photobiol.* **36**, 689–697 (1982)

Levine, L. D. and Ward, W. W. Isolation and characterization of a photoprotein, 'phialidin' and a spectrally unique green-fluorescent protein from the bioluminescent jellyfish *Phialidium gregarium*. *Comp. Biochem. Phys.* **72**, 77–86 (1982)

Lloyd, J. E. and Wing, S. R. Nocturnal aerial predation of fireflies by light-seeking fireflies. *Science* **222**, 634–635 (1983)

McCapra, F. and Hart, R. The origins of marine bioluminescence. *Nature* **286**, 660–661 (1980)

Merenyi, G., Lind, J. and Eriksen, T. E. The reactivity of superoxide (O_2^{-}) and its ability to induce chemiluminescence with luminol. *Photochem. Photobiol.* **41**, 203–208 (1985)

Posner, G. H., Lever, J. R., Miura, K., Lisek, C., Seliger, H. H. and Thompson, A. A chemiluminescent probe specific for singlet oxygen. *Biochem. Biophys. Res. Commun.* **123**, 869–873 (1984)

Roda, A., Girotti, S., Ghini, S., Grigolo, B., Carrea, G. and Bovara, R. Continuous-flow determination of primary bile acids, by bioluminescence, with use of nylon-immobilized bacterial enzymes. *Clinic. Chem.* **30**, 206–210 (1984)

Salin, M. L. and Bridges, S. M. Chemiluminescence in soybean root tissue: effect of various substrates and inhibitors. *Photobiochem. Photobiophys.* **6**, 57–64 (1983)

Seliger, H. H., Lall, A. B., Lloyd, J. E. and Biggley, W. H. The colors of firefly bioluminescence. I. Optimization model. *Photochem. Photobiol.* **36**, 673–680 (1982)

Shimomura, O. and Shimomura, A. Effect of calcium chelators on the Ca^{2+}-dependent luminescence of aequorin. *Biochem. J.* **221**, 907–910 (1984)

Sweeney, B. M., Fork, D. C. and Satoh, K. Stimulation of bioluminescence in dinoflagellates by red light. *Photochem. Photobiol.* **37,** 457–465 (1983)

Tanaka, K. and Ishikawa, E. Highly sensitive bioluminescent assay of dehydrogenases using NAD(P)H: FMN oxidoreductase and luciferase from *Photobacterium fischeri. Analyt. Lett.* **17,** 2025–2034 (1984)

Wier, W. G., Kort, A. A., Stern, M. D., Lakatta, E. G. and Marban, E. Cellular calcium fluctuations in mammalian heart: Direct evidence from noise analysis of aequorin signals in Purkinje fibers. *Proc. Natl. Acad. Sci. USA* **80,** 7367–7371 (1983)

PART III

Photobiology

The general photochemical principles discussed in Part II also apply to processes *in vivo*. However, the problems are more complicated: in sufficient dilution there are no interactions between molecules, a factor which cannot be neglected in biological systems. Photoreceptor molecules interact with similar or different neighboring molecules. The altered photochemical properties are reflected by changes in the absorption spectra. *In vivo* the absorption maxima of molecules are often shifted by many nm as compared to those in solution.

An additional complication is that primary photochemical reactions are sometimes difficult to separate from the subsequent processes. For example it is not easy to determine whether the conformational change of a protein is the direct result of the excitation by light, or whether it is due to a change in the pK or due to a proton translocation.

A

Light absorption for energy fixation

Absorption of radiation by photoreceptor pigments can serve one of two principally different processes. On the one hand it can induce, regulate or control movement or developmental processes. In this case a considerable amplification of the absorbed energy is often necessary (see Part III B). On the other hand the absorbed quanta can be used as an energy source. These anabolic reactions are performed by green plants and some bacteria. Most of these organisms use chlorophylls or related cyclic porphyrin ring systems for photosynthetic energy fixation. However, recently organisms have been found, and extensively studied, which transform light energy into chemical energy by a totally different mechanism. The best known example is *Halobacterium halobium*.

8 *Halobacterium*

The genus *Halobacterium* includes a number of morphologically and physiologically similar species which are adapted to a very special habitat with high or even saturated salt concentrations as found in the Dead Sea, in the Great Salt Lake in Utah or in salinas. The cells grow optimally in 3.5 to 5 M NaCl solutions and they lyse in concentrations lower than 3 M. The whole biochemical apparatus is adapted to a high salt concentration. Most enzymes, as well as the ribosomes and the cell wall, have their activity maximum above a concentration of 2 M and become unstable when the salt concentration of the medium is reduced.

The high external NaCl concentration is balanced by an equally high internal KCl concentration. Halobacteria are characterized by a number of other remarkable features. Though they are typical prokaryotes they lack the typical murein layer found in the cell wall of other bacteria; they also lack the widely distributed lipoproteins. Instead, their cell wall consists of a single glycoprotein which resembles one found on the cell surface of eukaryotic cells. It has a molecular weight of about 200,000 and contains about 10 to 12% carbohydrates, in the form of hexoses, amino sugars and uronic acids bound to the protein by O- and N-glycosidic bonds. The membrane also differs from that of other bacteria: instead of the common fatty acid glycerol esters it contains isoprenoid glycerol ethers.

A number of other biochemical properties, such as the chemical structure of ferredoxin, details in the protein biosynthesis and the amino acid sequence of the ribosomal protein also indicate a closer relationship to eukaryotes than to other bacteria. However, they are true prokaryotes, since they lack organelles such as nucleus and endoplasmatic membranes, and they swim using typical bacterial flagellae. They seem to be closely related to the archaic group of methanobacteria (archaebacteria).

The remarkable fact that halobacteria harvest light energy without using chlorophyll in contrast to all other photosynthetically active organisms could indicate that these organisms have developed during an early phase of evolution. One could speculate that they have developed a light energy fixation process before chlorophyll was 'invented'. Later on in evolution this system was replaced by the more effective, but also more complicated, photosynthesis.

Compared to the structural and biochemical apparatus of photosynthetic organisms the composition of the energy-fixing structures of *Halobacterium* is rather uniform and simple. When the cells grow at a reduced oxygen partial pressure, patches of the so-called purple membrane are developed in the cytoplasmic membrane (Fig. 8.1). The purple membrane consists of a uniform protein with a molecular weight of about 26,000 arranged in a quasi-crystalline hexagonal pattern. This protein is similar to the opsins and carries retinal as a chromophoric group, as do the visual pigments of animals. Therefore it has been called bacteriorhodopsin by Stoeckenius, in whose laboratory the biochemical analyses were carried out.

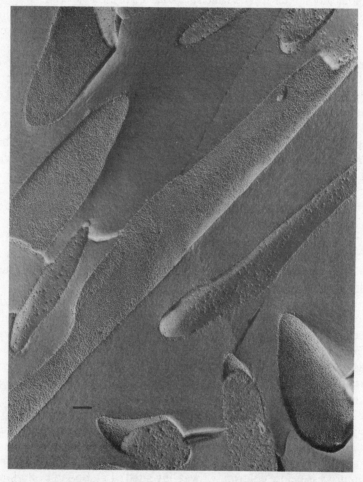

Fig. 8.1. Electron micrograph of a freeze–fracture preparation of the membrane of *Halobacterium halobium* with patches of the purple membrane. The scale represents about 0.2 μm (original photograph courtesy of J. Usukura, Tokyo).

Retinal is the aldehyde of vitamin A, which is a β-carotene split in half (cf. Fig. 8.4). The isolated retinal does not absorb in the visible range but in a broad band near 380 nm. The chromophoric group is linked to the amino acid lysine forming a Schiff's base, and is hidden in a hydrophobic pouch inside the protein. By this arrangement the absorption maximum is shifted to about 570 nm (Fig. 8.2). Electron optical investigations and X-ray analysis have shown that bacteriorhodopsin consists of seven protein helical domains which span the membrane perpendicularly (Fig. 8.3).

The remarkable similarity of bacteriorhodopsin to the visual pigments of animals could suggest that this pigment plays a role in photoorientation of halobacteria. In fact, Hildebrand found a photophobic reaction, the action spectrum of which corresponds to the absorption spectrum of bacteriorhodopsin. However, it has recently been demonstrated that the phobic response is mediated by a similar, related, pigment.

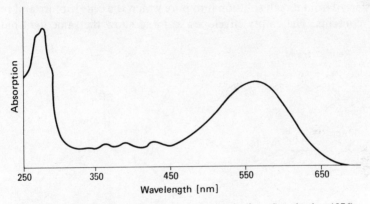

Fig. 8.2. Absorption spectrum of bacteriorhodopsin (from Stoeckenius, 1976).

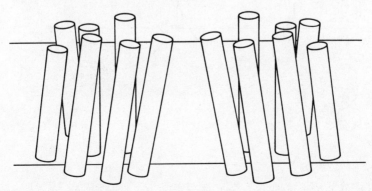

Fig. 8.3. Hypothetical structure of bacteriorhodopsin with seven helical domains spanning the membrane (from Jost and Packer, modified).

Another observation initiated the clarification of an interesting reaction: irradiation of the bacterium induced an increased ATP synthesis even though respiration was inhibited by light. The light energy-fixing process has been largely revealed by the combined efforts of a number of laboratories.

Absorption of a quantum of light induces a complex reaction cycle which has been analyzed *in vitro* (Fig. 8.4). During this cycle bacteriorhodopsin is bleached and transformed into an intermediate absorbing at 412 nm. The original form is regenerated *in vivo* within a few milliseconds, but by adding diethylether the intermediate can be conserved. During the reaction cycle bacteriorhodopsin occurs as an all-*trans* isomer and a 13-*cis* isomer.

The process described so far does not explain how light energy can be converted into chemical energy. The answer to this question was found when the pH was monitored as a function of irradiation. Oesterhelt found an acidification of the medium in light which indicated a proton extrusion out of the cell. When *Halobacterium* is alternately frozen and thawed, or when it is transferred from its salt solution into pure water, the cells rupture and release their contents. The empty envelopes seal and show the same light-induced

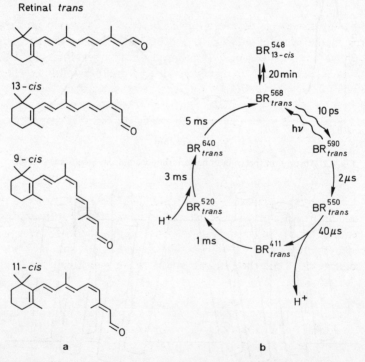

Fig. 8.4 (a) Retinal in all-*trans* and 13-*cis* form. (b) Light-induced reaction cycle of bacteriorhodopsin BR with the reaction constants and the absorption maxima (in nm) of the intermediates (from Hildebrand).

vectorial proton transport as the whole cells, provided they contain purple membrane patches. This experiment indicated that the protein forming the purple membrane is a light-dependent proton pump which generates a proton gradient in light. This reaction can even be induced in lipid vesicles into which bacteriorhodopsin has been incorporated.

The proton gradient can be demonstrated by adding a radioactively labelled lipophilic cation (triphenyl methyl phosphonium$^+$ bromide$^-$) which can easily penetrate biomembranes. For each extruded proton, one TPMP$^+$ ion enters the cell to compensate the negative charge. In the cell it can be detected due to its radioactive label. With this technique the size of the proton gradient can be quantified.

Since the transported protons carry an electric charge an electric potential difference $\Delta\psi$ is generated in addition to the pH gradient; both form the electrochemical gradient $\tilde{\mu}_{H^+}$

$$\tilde{\mu}_{H^+} = R\,\frac{I}{F}\,\Delta pH + \Delta\psi.$$

The energy stored in the proton gradient can be converted into chemical energy in the form of ATP by a process described by the chemiosmotic hypothesis for which Mitchell received the Nobel Prize in 1978 (see Section 6.3.5). According to this theory, the protons previously extruded from the cell follow the electrical gradient and re-enter the cell through a membrane-bound enzyme (ATP synthase). The proton flux drives the ATP synthesis like a water current drives a water mill. The ATP synthase can also operate in reverse and extrude protons from the cell's interior at the expenditure of ATP (Fig. 8.5). By this mechanism the cell can convert the proton gradient into ATP and vice-versa according to the circumstances.

In addition, *Halobacterium* has a normal respiratory chain and breaks down organic substances in the presence of oxygen. The energy released during this process is also stored in the form of a proton gradient and subsequently converted into ATP as well as the light-induced proton gradient. The purple membrane is only synthesized in light and at low oxygen concentrations. Thus the cell can adapt to the prevailing conditions and switch from the oxidation of organic substances to the utilization of solar energy.

Biomembranes are not very good barriers for protons over long periods of time, so that the generated proton gradient breaks down. The passive permeability reduces the effectivity of the energy converter. Therefore the primary proton gradient is rapidly converted into a potassium or sodium gradient by means of ion exchangers in the membrane which allow a stoichiometric antiport of protons verses potassium or sodium. In turn, the

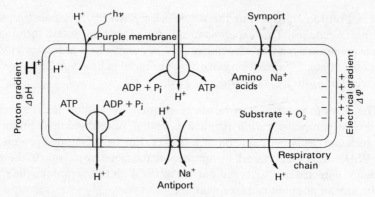

Fig. 8:5. Ion transport mechanisms in *Halobacterium halobium*. In light protons are transported through the purple membrane and in darkness by the respiration chain which builds up a pH gradient and an electrical gradient. A proton influx can drive an antiport of Na$^+$ ions or an ATP synthesis. The influx of Na$^+$ ions allows a symport influx of amino acids (from Hildebrand and Stoeckenius, modified).

sodium gradient can be used in a symport reaction to take up amino acids such as serine or asparagic acid into the cell.

8.1   BIBLIOGRAPHY

Textbooks and review articles

Dencher, N. A. The five retinal–protein pigments of halobacteria: bacteriorhodopsin, halorhodopsin, P 565, P 370, and slow-cycling rhodopsin. *Photochem. Photobiol.* **38**, 753–767 (1983)

Dencher, N. A. Photosensory function of retinal–protein pigments in *Halobacterium halobium*. In: Bolis, C. L., Helmreich, E. J. M. and Passow, H. (eds) *Information and Energy Transduction in Biological Membranes*. Alan, R. Liss, New York, pp. 231–236 (1984)

Hildebrand, E. and Schimz, A. Sensory transduction in *Halobacterium*. In: Colombetti, G., Lenci, F. and Song, P.-S. (eds) *Sensory Perception and Transduction in Aneural Organisms*. Plenum Press, New York and London, pp. 93–111 (1985)

Honig, B. Photochemical charge separation and active transport in the purple membrane. In: Slayman, C. L. (ed.) *Current Topics in Membranes and Transport*. Academic Press, New York (1982)

Keszthelyi, L. Intramolecular charge shifts during the photoreaction cycle of bacteriorhodopsin. In: Bolis, C. L., Helmreich, E. J. M. and Passow, H. (eds) *Information and Energy Transduction in Biological Membranes*. Alan R. Liss, New York, pp. 51–71 (1984)

Kushner, D. J. The halobacteriaceae. In: Woese, C. R. and Wolfe, R. S. (eds) *Bacteria: A Treatise on Structure and Function*. Vol. 8. *Archaebacteria*. Academic Press, Orlando, pp. 171–214 (1985)

Lanyi, J. K. Nature of the principal photointermediate of halorhodopsin. *Biochem. Biophys. Res. Commun.* **122**, 91–96 (1984)

Lozier, R. H. and Parodi, L. A. Bacteriorhodopsin: photocycle and stoichiometry. In: Bolis, C. L., Helmreich, E. J. M. and Passow, H. (eds) *Information and Energy Transduction in Biological Membranes*. Alan R. Liss, New York, pp. 39–50 (1984)

Neugebauer, D.-C., Zingsheim, H.-P. and Oesterhelt, D. Biogenesis of purple membrane in halobacteria. *Meth. Enzymol.* **97**, 218–226 (1983)

Packer, L. (ed.) Biomembranes. Pt. I. Visual pigments and purple membranes. In: *Methods in Enzymology.* Academic Press, New York (1982)

Schulten, K., Schulten, Z. and Tavan, P. An isomerization model for the pump cycle of bacteriorhodopsin. In: Bolis, C. L., Helmreich, E. J. M. and Passow, H. (eds) *Information and Energy Transduction in Biological Membranes.* Alan R. Liss, New York, pp. 113–131 (1984)

Spudich, J. L. Genetic demonstration of a sensory rhodopsin in bacteria. In: Bolis, C. L., Helmreich, E. J. M. and Passow, H. (eds) *Information and Energy Transduction in Biological Membranes.* Alan R. Liss, New York, pp. 221–229 (1984)

Spudich, J. Color-sensing by phototactic *Halobacterium halobium.* In: Colombetti, G., Lenci, F. and Song, P.-S. (eds) *Sensory Perception and Transduction in Aneural Organisms.* Plenum Press, New York and London, pp. 113–118 (1985)

Stoeckenius, W. The purple membrane of salt loving bacteria. *Sci. Amer.* **234,** 38–46 (1976)

Wagner, G. Blue light effects in halobacteria. In: Senger, H. (ed.) *Blue Light Effects in Biological Systems.* Springer-Verlag, Berlin, Heidelberg, New York and Tokyo, pp. 48–54 (1984)

Further reading

Abdulaev, N. G., Kiselev, A. V., Ovchinnikov, Y. A., Drachev, L. A., Kaulen, A. D. and Skulachev, V. P. C-terminal region does not affect proton translocating activity of bacteriorhodopsin. *Biol. Membrany* **2,** 453–459 (1985)

Ahl, P. L. and Cone, R. A. Light activates rotations of bacteriorhodopsin in the purple membrane. *Biophys. J.* **45,** 1039–1049 (1984)

Alshuth, T., Stockburger, M., Hegemann, P. and Oesterhelt, D. Structure of the retinal chromophore in halorhodopsin. A resonance Raman study. *FEBS Lett.* **179,** 55–59 (1985)

Bamberg, E., Hegemann, P. and Oesterhelt, D. Reconstitution of the light-driven electrogenic ion pump halorhodopsin in black lipid membranes. *Biochim. Biophys. Acta* **773,** 53–60 (1984)

Bamberg, E., Hegemann, P. and Oesterhelt, D. The chromoprotein of halorhodopsin is the light-driven electrogenic chloride pump in *Halobacterium halobium. Biochem.* **23,** 6216–6221 (1984)

Beach, J. M. and Fager, R. S. Evidence for branching in the photocycle of bacteriorhodopsin and concentration changes of late intermediate forms. *Photochem. Photobiol.* **41,** 557–562 (1985)

Bogomolni, R. A., Taylor, M. E. and Stoeckenius, W. Reconstitution of purified halorhodopsin. *Proc. Natl. Acad. Sci.* **81,** 5408–5411 (1984)

Dencher, N. A., Kohl, K. D. and Heyn, M. P. Photochemical cycle and light–dark adaptation of monomeric and aggregated bacteriorhodopsin in various lipid environments. *Biochem.* **22,** 1323–1334 (1983)

Drachev, A. L., Drachev, L. A., Evstigneeva, R. P., Kaulen, A. D., Lazarova, C. R., Laikhter, A. L., Mitsner, B. I., Skulachev, V. P., Khitrina, L. V. and Chekulageva, L. N. Electrogenic stages in the photocycle of bacteriorhodopsins with modified retinals. *Biol. Membrany* **1,** 1125–1142 (1984)

Drachev, L. A., Dracheva, S. M., Samuilov, V. D., Semenov, A. Y. and Skulachev, V. P. Photoelectric effects in bacterial chromatophores. Comparison of spectral and direct electrometric methods. *Biochim. Biophys. Acta* **767,** 257–262 (1984)

Drachev, L. A., Kaulen, A. D. and Skulachev, V. P. Correlation of photochemical cycle, H^+ release and uptake, and electric events in bacteriorhodopsin. *FEBS Lett.* **178,** 331–335 (1984)

Draheim, J. E. and Cassim, J. Y. Large scale global structural changes of the purple membrane during the photocycle. *Biophys. J.* **47,** 497–507 (1985)

Druckmann, S., Renthal, R., Ottolenghi, M. and Stoeckenius, W. The radiolytic reduction of the Schiff base in bacteriorhodopsin. *Photochem. Photobiol.* **40,** 647–651 (1984)

Efremov, R. G. and Nabiev, I. R. Microenvironment of the retinal chromophore in bacteriorhodopsin. *Biol. Membrany* **2,** 460–469 (1985)

Ehrenberg, B., Meiri, Z. and Loew, L. M. A microsecond kinetic study of the photogenerated membrane potential of bacteriorhodopsin with a fast responding dye. *Photochem. Photobiol.* **39,** 199–205 (1984)

Harbison, G. S., Smith, S. O., Pardeon, J. A., Winkel, C., Lugtenburg, J., Herzfeld, J., Mathies, R. and Griffin, R. G. Dark-adapted bacteriorhodopsin contains 13-*cis*, 15-*syn* and all-*trans*, 15-anti retinal Schiff bases. *Proc. Natl. Acad. Sci.* **81,** 1706–1709 (1984)

Hazemoto, N., Kamo, N., Kobatake, Y., Tsuda, M. and Terayama, Y. Effect of salt on photocycle and ion-pumping of halorhodopsin and third rhodopsinlike pigment of *Halobacterium halobium*. *Biophys. J.* **45**, 1073–1077 (1984)

Henderson, R. and Unwin, P. N. T. Three dimensional model of purple membrane obtained by electron microscopy. *Nature* **257**, 28–32 (1975)

Jubb, J. S., Worcester, D. L., Crespi, H. L. and Zaccai, G. Retinal location in purple membrane of *Halobacterium halobium*: a neutron diffraction study of membranes labelled *in vivo* with deuterated retinal. *EMBO J.* **3**, 1455–1461 (1984)

Kimura, Y., Ikegami, A. and Stockenius, W. Salt and pH-dependent changes of the purple membrane absorption spectrum. *Photochem. Photobiol.* **40**, 641–646 (1984)

Lanyi, J. K. Physicochemical aspects of salt-dependence in halobacteria. *Life Sci. Rep.* **13**, 93–107 (1979)

Lanyi, J. K. Light-dependent *trans* to *cis* isomerization of the retinal in halorhodopsin. *FEBS Lett.* **175**, 337–342 (1984)

Li, Q.-Q., Govindjee, R. and Ebery, T. G. A correlation between proton pumping and the bacteriorhodopsin photocycle. *Proc. Natl. Acad. Sci.* **81**, 7079–7082 (1984)

Ogurusu, T., Maeda, A. and Yoshizawa, T. Absorption spectral properties of purified halorhodopsin. *J. Biochem.* **95**, 1073–1082 (1984)

Ovchinnikov, Y. A., Abdulaev, N. G., Vasilov, R. G., Vturina, I. Y., Kuryatov, A. B. and Kiselev, A. V. The antigenic structure and topography of bacteriorhodopsin in purple membranes as determined by interaction with monoclonal antibodies. *FEBS Lett.* **179**, 343–350 (1985)

Parodi, L. A., Lozier, R. H., Bhattacharjee, S. M. and Nagle, J. F. Testing kinetic models for the bacteriorhodopsin photocycle- II. Inclusion of an O to M backreaction. *Photochem. Photobiol.* **40**, 501–512 (1984)

Renard, M., Thirion, P. and Delmelle, M. Photoacoustic spectroscopy of bacteriorhodopsin photocycle. *Biophys. J.* **44**, 211–218 (1983)

Rothschild, K. J., Marrero, H., Braiman, M. and Mathies, R. L. Primary photochemistry of bacteriorhodopsin: comparison of Fourier transform infrared difference spectra with resonance Raman spectra. *Photochem. Photobiol.* **40**, 675–679 (1984)

Seltzer, S. and Ehrenson, S. The purple membrane proton pump: A mechanistic proposal for the source of the second proton. *Photochem. Photobiol.* **39**, 207–211 (1984)

Shichida, Y., Matuoka, S., Hidaka, Y. and Yoshizawa, T. Absorption spectra of intermediates of bacteriorhodopsin measured by laser photolysis at room temperatures. *Biochim. Biophys. Acta* **723**, 240–246 (1983)

Smith, S. O., Marvin, M. J., Bogomolni, R. A. and Mathies, R. A. Structure of the retinal chromophore in the hR_{578} form of halorhodopsin. *J. Biol. Chem.* **259**, 12326–12329 (1984)

Spudich, E. N. and Spudich, J. L. Biochemical characterization of halorhodopsin in native membranes. *J. Biol. Chem.* **260**, 1208–1212 (1985)

Spudich, J. L. and Bogomolni, R. A. Spectroscopic discrimination of the three rhodopsinlike pigments in *Halobacterium halobium* membranes. *Biophys. J.* **43**, 243–246 (1983)

Spudich, J. L. and Bogomolni, R. A. Mechanism of colour discrimination by a bacterial sensory rhodopsin. *Nature* **312**, 509–513 (1984)

Sugiyama, Y. and Mukohata, Y. Isolation and characterization of halorhodopsin from *Halobacterium halobium*. *J. Biochem.* **96**, 413–420 (1984)

Takahashi, T., Mochizuki, Y., Kamo, N. and Kobatake, Y. Evidence that the long-lifetime photointermediate of S-rhodopsin is a receptor for negative phototaxis in *Halobacterium halobium*. *Biochem. Biophys. Res. Commun.* **127**, 99–105 (1985)

Takahashi, T., Tomioka, H., Kamo, N. and Kobatake, Y. A photosystem other than PS370 also mediates the negative phototaxis of *Halobacterium halobium*. *FEMS Microbiol. Lett.* **28**, 161–164 (1985)

Takeuchi, Y., Ohno, K., Yoshida, M. and Nagano, K. Light-induced proton dissociation and association in bacteriorhodopsin. *Photochem. Photobiol.* **33**, 587–592 (1981)

Taylor, M. E., Bogomolni, R. A. and Weber, H. J. Purification of photochemically active halorhodopsin. *Proc. Natl. Acad. Sci. USA* **80**, 6172–6176 (1983)

Tomioka, H., Kamo, N., Takahashi, T. and Kobatake, Y. Photochemical intermediate of third rhodopsin-like pigment in *Halobacterium halobium* by simultaneous illumination with red and blue light. *Biochem. Biophys. Res. Commun.* **123**, 989–994 (1984)

Tsuda, M., Nelson, B., Chang, C.-H., Govindjee, and Ebrey, T. G. Characterization of the chromophore of the third rhodopsin-like pigment of *Halobacterium halobium* and its photoproduct. *Biophys. J.* **47**, 721–724 (1985)

Wolber, P. K. and Stoeckenius, W. Retinal migration during dark reduction of bacteriorhodopsin. *Proc. Natl. Acad. Sci.* **81**, 2303–2307 (1984)

9 Photosynthesis

9.1 INTRODUCTION

The radiation energy of visible light is used in the photosynthesis of green plants to synthesize organic compounds from the simple inorganic molecules carbon dioxide and water. Oxygen is photolytically cleaved from water. The radiation energy is stored in carbohydrates in the form of Gibbs free energy $\Delta G°$. The endergonic reaction can be summarized by the following equation

$$6\,CO_2 + 12\,H_2O \rightarrow C_6H_{12}O_6 + 6\,H_2O + 6\,O_2 \quad \Delta G° = 2868\,kJ/mol\,C_6H_{12}O_6.$$

Chemically seen, this process is a reduction of CO_2 since the inverse reaction is the oxidation of carbohydrates. CO_2 is reduced by a reductant in the form of NADPH and for thermodynamic reasons ATP is also required.

$$CO_2 + 2\,NADPH + 3\,ATP^{4-} + H_2O \rightarrow$$
$$[CH_2O] + 2\,NADP^+ + 3\,ADP^{3-} + 3\,HPO_4^{2-} + H^+$$

(For reasons of simplification we use NADP and NADPH instead of $NADP^+$ and $NADPH + H^+$). In green plants and algae, NADPH and ATP for carbohydrate synthesis are produced exclusively in specific organelles, the chloroplasts. Each chloroplast is enclosed by a double membrane which separates its interior from the cellular cytoplasm (Fig. 9.1). The inner phase is called stroma and contains all the necessary enzymes for CO_2 fixation and carbohydrate synthesis. Within the stroma we find thylakoids which are membranous structures where NADPH and ATP are produced in light dependent reactions. The biochemical reactions of CO_2 fixation and reduction, as well as the subsequent reactions of the products, occur in the stroma independent of light (Fig. 9.2a).

The photosynthetic production of NADPH is a redox reaction in which NADP is reduced upon uptake of two electrons. The electron is freed during the splitting of water, which also results in the production of oxygen. The redox pair NADPH/NADP has a high free energy or – electrochemically speaking – a negative redox potential. In contrast, the pair H_2O/O_2 has only a small free energy and a positive redox potential. Substances with a more negative redox potential can reduce substances with a more positive redox

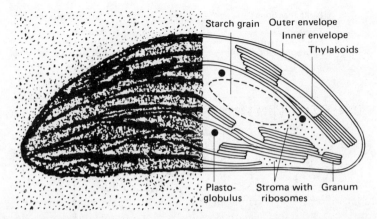

Fig. 9.1. Chloroplast of a higher plant. Left: electron optical photograph; right: schematic cross-section.

potential. This process occurs, for instance, during respiration, a biological oxidation, in which NADH reduces oxygen to produce water. In respiration the energy difference between NADH and water is utilized to produce ATP. In photosynthesis the electromagnetic energy of light is used to cover the potential difference between water and NADPH. For this purpose electrons need to be lifted from a lower to a higher energy level against a thermodynamic gradient. In photosynthesis this is effected by two light reactions in sequence using two specialized chlorophylls, P680 and P700, which convert the electromagnetic energy into chemical energy. These chlorophylls form the reaction centers of photosystems I (PSI) and II (PSII) and are adjacent to the redox systems which connect the two reaction centers in a common chain (Fig. 9.2b). Chlorophylls and redox systems form an electron transport chain which starts with the water splitting system freeing the electrons and protons for the chain allied to oxygen evolution.

$$2\ H_2O \rightarrow O_2 + 4e^- + 4H^+$$

The other end of the electron transport chain – which is activated by the energy of the absorbed quanta – is defined by the reduction of NADP by electrons from the chain and protons from the stroma.

$$NADP^+ + 2e^- + 2H^+ \rightarrow NADPH + H^+$$

The photolysis of water occurs inside the closed vesicular thylakoids and causes an acidification of the interior due to the proton release. NADP reduction occurs at the outside of the thylakoids and dissipates protons. Vectorial electron transport through the thylakoid membrane thus results in a proton gradient with a pH difference of 3 to 4 pH units between the thylakoid

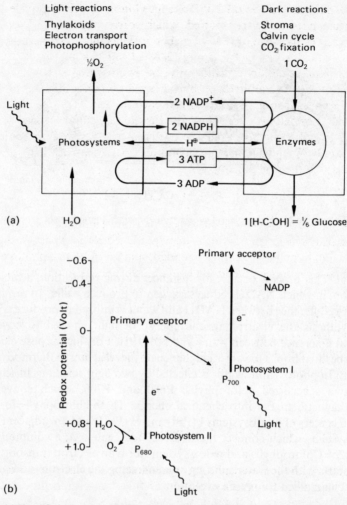

Fig. 9.2. (a) Simplified schematic diagram of photosynthesis (from Lichtenthaler and Pfister, 1978, modified). (b) Z-diagram of photosynthetic light reactions indicating redox potentials.

inside and the stroma. The resulting electrochemical potential difference, also called proton motive force, is used by an ATP-synthase to produce ATP.

Thus the photochemical primary reactions can be seen as an electron transport between water with a redox potential of $+0.82$ V (at pH 7.25 and $0°C$) to NADP with a redox potential of -0.32 V. The generation of the proton gradient for ATP synthesis is closely coupled with the electron transport. Since NADP is reduced by the more negative redox system ferredoxin ($E'_0 = -0.43$ V) the electrons have to be pushed 'uphill' against a potential difference of about 1.2 V. To reduce 1 mole of carbon dioxide 4 moles of electrons need to be transported, which corresponds to 4.8 V or 478 kJ ($=114$ kcal) in the form of radiation energy. Thus the production of 1 mole glucose (reduction of 6 CO_2) requires 2868 kJ ($=685$ kcal).

The overall equation is valid only for photosynthetic eukaryotes and cyanobacteria ($=$ cyanophyceae $=$ blue–green algae) since only they use water as an electron donor. Sulfur purple bacteria use H_2S or other sulfur compounds instead of water. Therefore they release sulfur (instead of water) which is stored in globuli inside the bacterial cell. Van Niel recognized the similarities between bacterial and plant photosynthesis as early as 1931, and he formulated the general equation

$$CO_2 + 2\,H_2A + light \rightarrow [CH_2O] + H_2O + 2A$$

In addition to the biochemical differences there are a number of structural and functional characteristics which distinguish between photosynthesis in bacteria and in plants. In contrast to algae and green plants, photosynthetic bacteria and cyanobacteria do not contain chloroplasts, and their thylakoids are distributed as closed compartments within the cytoplasm. The free bacterial thylakoids have a different substructure from the chloroplast thylakoids. In contrast to plant photosynthesis, bacterial photosynthesis uses only one light reaction. However, some recent investigations indicated that higher plants also need only one light reaction to reduce NADP, but this hypothesis has not yet been substantiated.

9.2 COMPOSITION AND STRUCTURE OF CHLOROPLASTS

Chloroplasts, the photosynthetic organelles, are usually lens-shaped in mosses, ferns and seed plants while they are heteromorph in algae. In higher plants they are usually 4–8 μm in diameter and 2–3 μm thick. Their number per cell varies greatly among plant species. The assimilatory parenchyma (palisade and spongy parenchyma) of leaves contains a specially large number of chloroplasts. In spinach leaves, for instance, 300–400 chloroplasts have been found per palisade parenchyma cell and 200–300 per sponge parenchyma cell. The dry weight of spinach chloroplasts is about 9×10^{-12} g and that of corn chloroplasts about 38×10^{-12} g. Proteins amount to 60% of the dry weight in spinach chloroplasts, lipids to about 20% and nucleic acids to up to 4%. Inorganic ions (Mg^{2+}, Mn^{2+}, Ca^{2+}, Na^+, K^+, Cl^-) can also add up to about 4% of the dry weight (Table 9.1A). The amount of starch present varies. Among the lipids, chlorophylls, carotenoids and prenylquinones are important for functional reasons, and glyco- and phospholipids for structural

TABLE 9.1A   CHEMICAL COMPOSITION OF INTACT SPINACH CHLOROPLASTS
(FROM KIRK AND TILNEY-BASSET)

Component	Percentage of chloroplast (dry weight)	
Total protein	60	
water-insoluble protein		27
water-soluble protein		33
Total lipids	20	
chlorophylls		4.3
carotenoids		0.9
colorless lipids		14.8
RNA	1.7–3.5	
DNA	?	
Free amino acids	2.5	
Orthophospates	0.7	
Ions (K^+, Na^+, Mg^{2+}, Ca^{2+}, Cl^-)[a]	3.6	
Polysaccharides[b]	0.6	
Hexosamines	0.2	
Starch	variable	

[a]Data of *Beta vulgaris* chloroplasts
[b]Data of *Allium porrum* chloroplasts

TABLE 9.1B   LIPID COMPOSITION OF SPINACH CHLOROPLASTS
(BASED ON 100 MOL CHLOROPHYLL) (TEVINI, UNPUBLISHED)

Lipid	Mol	
Chlorophylls	100	
chlorophyll *a*		70
chlorophyll *b*		30
Carotenoids	39	
β-carotene		9
lutein		14
violaxanthin		10
neoxanthin		4
antheraxanthin		2
Quinones	11	
plastoquinone		4
plastohydroquinone		2
α-tocopherol		4
vitamin K_1		0.5
tocoquinone		0.5
Glycolipids	447	
monogalactosyldiglyceride		270
digalactosyldiglyceride		154
sulfolipid		23
Phospholipids	43	
phosphatidylglycerine		29
phosphatidylcholine		10
phosphatidylethanolamine		2
phosphatidylinosite		2

reasons. In young chloroplasts there are about 40 carotenoids, 10 prenylquinones, as well as about 500 glycolipids and 50 phospholipids per 100
chlorophyll molecules (Table 9.1B).

Chloroplasts are surrounded by a double envelope which encloses the
stroma (matrix). Within the stroma we find thylakoids, ribosomes, nucleoids
(DNA), plastoglobuli (lipid reservoirs) and starch grains (Fig. 9.1). Algae
possess pyrenoids which organize the deposition of reserve substances.

The thylakoids are closed membraneous systems which in mosses, ferns and
seed plants form stacks called grana thylakoids visible by light microscopy.
Irregular thylakoids extend from the grana into the stroma (stroma
thylakoids). The photosystems I and II are incorporated into the lipid matrix
of the thylakoids as pigment–protein particles of different size. Both
photosystems are connected by the cytochrome b_6–f complex which spans the
thylakoid membrane as an integral protein. The ATP-synthase is spatially
separated from this complex and forms another integral protein in the
thylakoid membrane. The photosystems are linked with peripheral proteins,
one of which is bound to the thylakoid interior and is responsible for water
splitting. The other is attached to the outside and catalyzes the NADP
reduction. The pigment–protein particles have been isolated from the
thylakoid membrane and analyzed using physical and biochemical techniques. Each photosystem consists of a central core complex (CC) and a light
harvesting complex (LHC). By the use of detergents the CC can be further split
into the pigmented reaction center subunits (RC) and unpigmented proteins.
Analogously, the LHC can be split into the light-harvesting pigment-proteins
(Section 9.3.1.1).

In the CC (also called photosystems in some thylakoid models) the
photochemical and biochemical responses occur which involve the chlorophylls and redox systems. We can regard the redox systems of the electron
transport chain as enzymes with the property of oxido-reductases, which are
membrane-bound in contrast to the stroma enzymes. This holds for the
cytochrome b_6–f complex and plastocyanin which connect the two photosystems in the thylakoid membrane. The ATP-synthases, the protein complexes
responsible for photophosphorylation, are found in regions where the
thylakoids are directly adjacent to the stroma. The arrangement of the various
particle complexes with their subunits is shown as a hypothetical thylakoid
model in Fig. 9.3.

Particular structures can also be observed by electron microscopy of freeze-
etched thylakoids. In order to preserve the structures the material is cut
while it is deep frozen. After sublimation of the surface ice the fractured
surface is sputtered with metal in vacuum, and the replica can be observed
electron-optically. It is difficult, however, to correlate the particles seen
with this technique with the functional protein complexes. Only in some

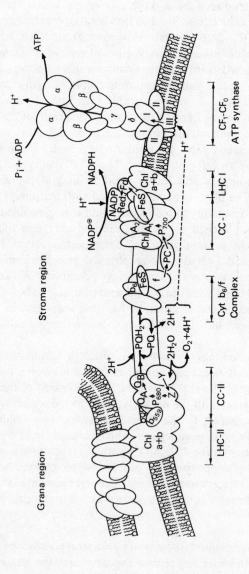

Fig. 9.3. Model of the thylakoid structure of higher plants and green algae. The individual peptides are shown as ovals or circles (from Kaplan and Arntzen, modified).

cases can the correlation between structure and function be demonstrated by a comparison of structure and size of the particles seen with the electron microscope with that of isolated functioning particles. For instance, electron-optical images show particles with a size of more than 14 nm, linked with photosystem II. They consist of 8 nm CC particles and a LHC of varying size. 15.5 nm particles could be identical with PSI particles. In stroma thylakoids additional particles have been found which could represent the ATP-synthase. Particles of similar size but loosely bound to the stroma thylakoids correspond to the ribulose-*bis*-phosphate (RubP) carboxylase, the enzyme involved in CO_2 fixation.

According to electron-optical and biochemical results the photosystems are quantitatively distributed differently in the grana and stroma thylakoids. In the densely packed grana thylakoids we find a high concentration of PSII particles, and in the stroma thylakoids mainly PSI particles. In addition, recent investigations have shown that the PSII particles are different in composition in grana and stroma thylakoids, but the functional significance is still obscure. Chlorophylls and carotenoids as components of the LHC play a central role in the photosystems since they absorb the radiation energy and funnel it to the reaction centers.

9.3   PHOTOSYNTHETICALLY ACTIVE PIGMENTS AND THEIR FUNCTIONS

Engelmann showed as early as 1882, that certain ranges of visible radiation are especially effective for photosynthetic oxygen production in filamentous green algae. He projected a simple prism spectrum onto an alga trichome and observed that aerotactic bacteria accumulated in the red, blue and violet spectral regions and concluded that the alga produces large amounts of oxygen in these spectral bands (Fig. 9.4a). With this technique Engelmann obtained a crude action spectrum of the relative quantum efficiency for photosynthetic oxygen production.

The photosynthetic action spectra agree fairly well with the total absorption spectra of the chlorophylls and carotenoids present in the chloroplasts, which indicates that these pigments are involved in the photosynthetic energy absorption (Fig. 9.4b,c). The photochemical primary processes of photosynthesis are, however, exclusively driven by specially bound chlorophyll a molecules with absorption maxima at 680 nm (P680) and 700 nm (P700). The bulk of the remaining chlorophylls and the carotenoids (and the phycobilins found in red algae, cyanobacteria and cryptophyceae) harvest the light energy and transfer it to the active chlorophylls. Due to the differences in pigment composition, the absorption spectra of red, green and brown algae, as well as

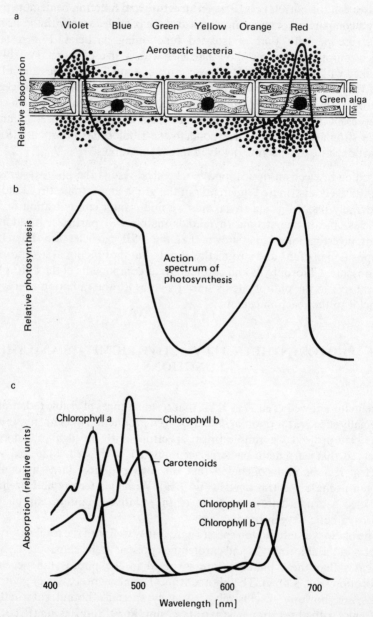

Fig. 9.4. (a) Aerotactic accumulation of bacteria around those parts of a green alga irradiated with red or blue light. (b) Action spectrum of photosynthesis and (c) absorption spectrum of chlorophyll *a* and *b*, and carotenoids in methanol (from Raven *et al.*, 1981, modified).

of cyanobacteria, differ (Fig. 9.5). Photosynthetic bacteria contain specific chlorophylls, the bacteriochlorophylls, in addition to carotenoids.

9.3.1   Chlorophylls

All photosynthetic organisms possess chlorophylls bound to protein complexes from which several chemically different forms can be extracted using organic solvents. In organisms with oxygen evolution chlorophyll a plays the major role. Willstätter and Stoll (in 1913 and 1918) as well as Fischer (in 1940) analyzed the chemical structure of chlorophyll as a tetrapyrrole with porphyrin ring structure and a magnesium central atom (Fig. 9.6). In addition to the pyrrole rings A–D chlorophyll has a cyclopentanone ring with a methyl ester. A phytol ester is linked by a propionic acid with pyrrole ring D. Chlorophyll b only differs from chlorophyll a in having a formyl group at C atom 7 instead of a methyl group. Other chlorophylls, such as chlorophyll c_1 and c_2, are found in brown algae and chlorophyll d in red algae.

Photoautotrophic bacteria possess bacteriochlorophyll a which differs from chlorophyll a in two features: it is hydrated in ring B and carries a vinyl group instead of the acetyl group on ring A. The green photobacterium, *Chlorobium*, has been found to contain the bacteriochlorophylls c, d and e, which have formerly been called *Chlorobium* chlorophylls. They contain farnesol instead of phytol attached to ring D. Bacteriochlorophyll b, which has been found for instance in *Rhodopseudomonas viridis*, probably carries geranylgeraniol instead of phytol.

The absorption of chlorophylls in the blue and red spectral range is caused by conjugated double bonds. While chlorophyll a has absorption maxima at 430 nm and 663 nm when dissolved in methanol, the maxima are shifted in chlorophyll b to 453 nm and 642 nm due to the formyl group in ring B (Fig. 9.5). Green plants usually have both chlorophyll a and b; the resulting broader absorption spectra allow a better utilization of the incident light energy. Some yellow–green algae, however, such as *Pleurochloris* (Eustigmatophyceae) contain only chlorophyll a. Removing the central magnesium atom by acid treatment results in brown pheophytins with absorption maxima slightly shifted toward shorter wavelengths compared with the corresponding chlorophylls.

Protochlorophyllide is a precursor in the chlorophyll biosynthesis; it has a double bond in ring D which is hydrated during photoreduction. Bacteriochlorophyll a has one double bond less than chlorophyll a (and two less than protochlorophyll) because of the saturated bond between C_7 and C_8 in ring B. The lower number of double bonds causes an absorption shift to longer wavelengths. Protochlorophyll absorbs in ether at 623 nm, chlorophyll a at

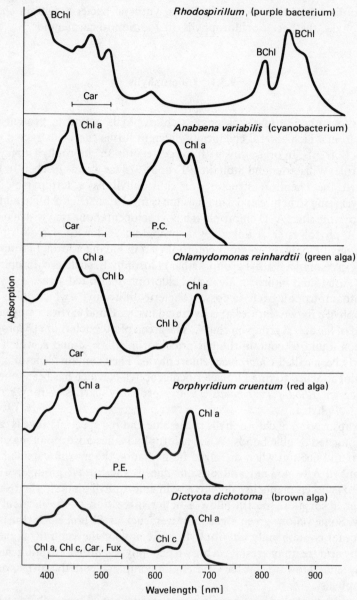

Fig. 9.5. Absorption spectra of purple bacteria, cyanobacteria, green, red and brown algae.
Chl = chlorophyll,   BChl = bacteriochlorophyll,   Car = carotenoids,   P.C. = phycocyanin,
P.E. = phycoerythrin, Fux = Fucoxanthin.

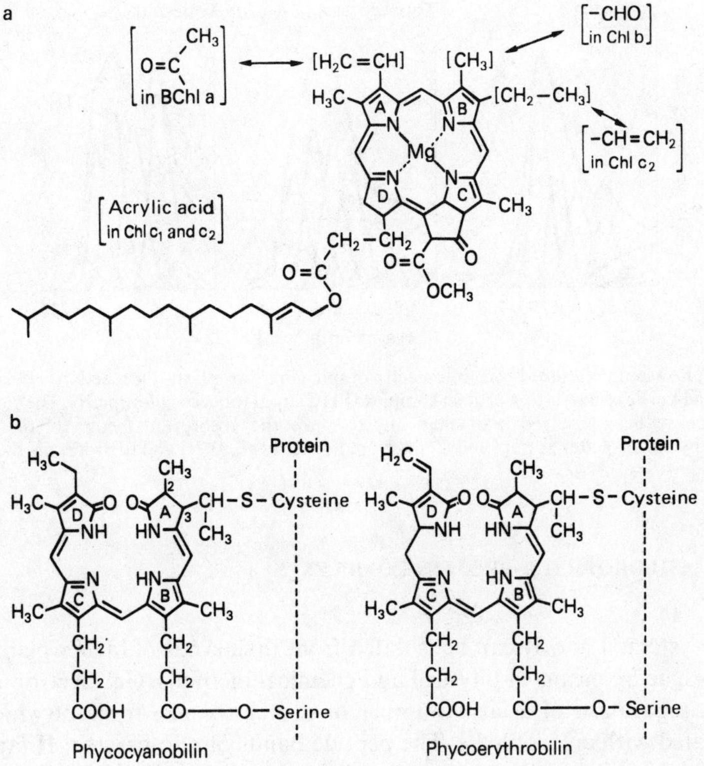

Fig. 9.6. (a) Chemical structures of chlorophylls a, b, c_1 and c_2 as well as bacteriochlorophyl a and of phycobilins (b).

663 nm and bacteriochlorophyll at 772 nm. Due to the hydrophilic porphyrin ring system and the lipophilic phytol, chlorophylls are strongly polarized, which facilitates their incorporation into membranes. *In vivo* chlorophylls are associated with proteins to form chlorophyll–protein complexes, which again causes a shift to longer wavelengths. At room temperature we find a broad absorption band centered around 675 nm. The analysis of a low-temperature spectrum by means of the mathematical method of deconvolution (see Chapter 4) reveals the presence of about six chlorophyll a forms with different absorption maxima between 660 nm and 706 nm. Figure 9.7 shows the deconvolution spectra for spinach chloroplasts and the core complex 1 isolated from tobacco chloroplasts (also called chlorophyll–protein complex 1 or CP1) as well as for the light-harvesting complex LHC. The different chlorophyll absorptions are thought to be due to interactions of the chlorophylls with lipids and proteins.

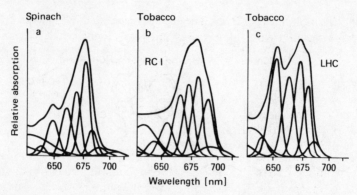

Fig. 9.7. Low-temperature absorption spectra of spinach chloroplasts, the reaction center protein (RCl) and the light-harvesting protein complex (LHC) from tobacco chloroplasts. The composite spectra have been resolved mathematically to show the absorption spectra of the various chlorophyll forms (after Stumpf and Conn, from French *et al.*, 1972, and Brown *et al.*, modified).

9.3.1.1 CHLOROPHYLL–PROTEIN COMPLEXES

Photosystems I and II can be isolated from thylakoids of higher plants and green algae by means of physical and chemical methods. Gel electrophoresis allows separation of a large number of protein bands, some of which are associated with chlorophylls. The peptide bands of photosystem II (approx. 10) include reaction center complex II peptides with a molecular weight of 34 kD, which in addition to 40–60 chlorophyll *a* molecules also contain the reaction center chlorophyll, P680. These peptides are missing in a *Chlamydomonas* and some barley mutants in which PSII activity is reduced or blocked. The LHCII is functionally associated with PSII and can be separated by gel electrophoresis into four to six peptides which contain chlorophylls *a* and *b* and carotenoids. In contrast, four other, smaller PSII polypeptides do not contain pigments. Three of them are attached peripherically to the thylakoid inner side, and are probably used for water splitting. The fourth protein, with a molecular weight of 32 kD, is associated with the thylakoid outer side and connects RCII with the intersystem electron transport chain. This protein has specific binding sites for herbicides, and is therefore called herbicide binding protein (HBP-32).

After detergent treatment photosystem I can be separated by gel electrophoresis into 13 bands which contain chlorophyll–protein complexes associated with the reaction center complex and the LHC. LHCI, the existence of which has long been doubted, can be separated into four polypeptides which together contain 40–80 chlorophylls. RCI (also called P700-chlorophyll *a* protein) has a higher molecular weight than RCII and contains 40–120 chlorophyll *a* molecules per P700. The peptides separated by gel electrophoresis after detergent treatment are associated to complexes *in vivo* with a molecular weight of about 110 kD, which is considerably higher than that of RCII.

It is still obscure how the isolated chlorophyll protein complexes are connected with the LHCs *in vivo* to form the photosynthetic units. A photosynthetic unit can be defined as a functioning complex of PSI and PSII. The size of this unit has often been described by the number of chlorophyll molecules it contains per electron transport chain (or per cytochrome *f*). Emerson and Arnold found a ratio of 600 chlorophylls per electron transport chain. *In vivo*, however, the number of chlorophylls per RC and LHC strongly depends on environmental factors, especially light. Therefore the size of the photosynthetic unit is, statistically seen, very variable. Figure 9.8 shows a hypothetical model in which 400 chlorophyll molecules are connected in a functioning unit.

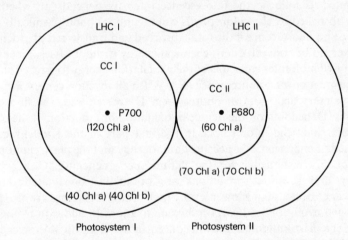

Fig. 9.8. Hypothetical model of the arrangement of chlorophyll protein complexes in a photosynthetic unit. CC = core complex, LHC I and II = light-harvesting complexes of photosystems I and II (from Stumpf).

9.3.1.2 CHLOROPHYLL FLUORESCENCE

Irradiation of chlorophyll *a* in solution with blue or red light causes a red fluorescence. The fluorescence emission spectrum is shifted toward longer wavelengths by about 7 nm relative to the absorption spectrum (Stokes shift, see Section 5.2.3 and Fig. 5.6). In functioning chloroplasts most of the incident energy is used photochemically. Therefore, *in vivo* we find only a small fluorescence yield (about 1%) in contrast to solutions in which photosynthesis does not take place. The fluorescence yield also increases in intact chloroplasts when photosynthesis is blocked by inhibitors such as diuron. In contrast to chlorophyll *a* solutions, intact chloroplasts show several fluorescence emission maxima which correspond to the various chlorophyll forms of photosystems I and II. At a temperature of $-196°C$ isolated chloroplasts show fluorescence emission maxima at 685 nm (F-685), 695 nm (F-695) and 735 nm (F-735). F-685 is associated with the antenna chlorophylls of LHCII, F-695 with the RC chlorophylls and F-735 with photosystem I. *In vivo*, virtually all the fluorescence is emitted by photosystem II.

After previous dark adaptation intact leaves or chloroplasts show characteristic, temporal chlorophyll fluorescence changes upon irradiation which were first observed by Kautsky and are called variable fluorescence. Switching the light on causes the chlorophyll fluorescence to rise within 1 s, from an initial level to a maximum from which it declines toward a lower steady state within several minutes (Fig. 9.9a). This induction curve can be interpreted as follows: the fluorescence intensity depends on whether the energy absorbed by the chlorophylls can be utilized photochemically. There would be no fluorescence if the total absorbed and transferred photochemical energy could be converted into chemical energy without losses. Since not all chlorophyll molecules can transfer their excitation energy to a reaction center we measure a ground fluorescence (F_0). When all reaction centers are closed, or all primary acceptors of photosystem II are reduced, the fluorescence increases. Depending on the degree of oxidation or reduction of the electron transport chain the fluorescence in the first part of the Kautsky curve is quenched or enhanced. The acceptor quinone Q_A and the plastoquinone pool are probably responsible for the fluorescence quenching. Only when all primary and secondary reactions are optimally coordinated does the fluorescence decrease to a lower level. Similar initial processes are also known for oxygen evolution and carbon dioxide fixation. In addition to the redox state of the plastoquinones, the ion concentration (e.g. Mg^{2+}) and especially the proton gradient, affect chlorophyll fluorescence. The ion effects may be due to conformational changes in the photochemical apparatus, especially the spatial distance between the LHC and the reaction center, which influences the energy transfer and thus the fluorescence yield.

Measurement of the variable chlorophyll fluorescence is an important tool in photosynthesis research, which can be used to analyze partial reactions of the electron transport chain and the site of inhibitor activity. This technique has been used successfully to reveal the mechanisms of PSII herbicides, such as ureas and triazines. Figure 9.9b shows the fast part of the fluorescence induction curve in pea chloroplasts under the effect of

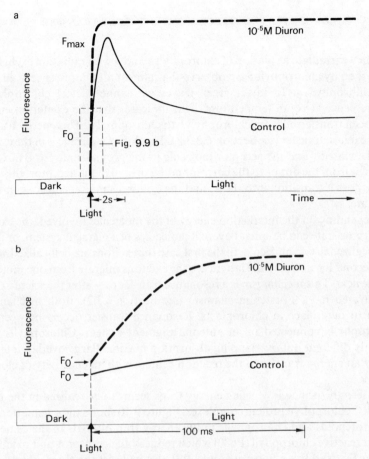

Fig. 9.9. (a) Schematic chlorophyll fluorescence induction curve in the seconds scale (from Govindjee and Papagiorgiou). (b) fluorescence induction curve in the milliseconds scale after irradiation of pea chloroplasts. Upper curves: fluorescence induction after inhibition of the electron transport chain by diuron. Excitation: blue light about 400 nm. Emission: red fluorescence >665 nm (redrawn from an original photograph by Dr. K. Pfister).

diuron. In contrast to the control, the fluorescence rapidly increases to a maximum since diuron inhibits the electron transport on the acceptor side of PSII (also see Fig. 9.20).

The measurement of fluorescence emission of, for instance, defective mutants indicates a lack of chlorophyll protein complexes *in vivo*. A *Chlamydomonas* mutant which lacks PSI activity also lacks the corresponding 735 nm fluorescence but not PSII fluorescence.

9.3.1.3 CHLOROPHYLL FLUORESCENCE AND ENERGY TRANSFER

When we irradiate a solution of chlorophyll a and b with radiation exclusively absorbed by chlorophyll b we observe that almost all fluorescence is emitted by chlorophyll a. In intact green leaves we cannot detect chlorophyll b fluorescence at room temperature. This indicates that the excitation energy has been transferred from chlorophyll b to chlorophyll a. This energy transfer is an exciton transfer (see Section 6.3.3.3). The donor molecule is in the excited singlet state S_1 and the acceptor molecule in the ground state S_0. By exciton transfer from the donor to the corresponding orbital of the acceptor, the latter undergoes a transition into the S_1 and the donor a simultaneous one into the S_0 state.

Depending on the interaction energy of the molecules involved, the excited electronic state can be spread over all molecules of a pigment system, or it can be localized in one molecule. In the first case the excitons are delocalized and in the second localized; in the latter case the excitons migrate from one molecule to the next. This mechanism is a resonance transfer or – after the scientist who discovered it – a Förster mechanism (see Section 6.3.2). Both mechanisms seem to take place in chloroplasts. Resonance transfer occurs between the chlorophylls connected in an antenna pigment system. Chlorophylls with slightly different but overlapping absorption maxima (large overlap integral) allow an energy transfer to the reaction centers with an effectivity of close to 100%.

In vivo the principle of using energy traps seems to be realized in the LHC and RC chlorophyll protein complexes. Figure 9.10 shows the schematic array of carotenoids and chlorophylls in the energy trap of PSII. In the center we find a reactive chlorophyll (P680) which reduces an acceptor A and oxidizes a donor D. In addition to carotenoids, PSI also includes the chlorophylls a_{662}, a_{670}, a_{677}, a_{684} and a_{692} as antenna pigments and P700 as the reaction center chlorophyll. In the LHC of higher plants and green algae, we mostly find chlorophyll b as accessory pigment, which transfers its energy without losses, and carotenoids which transfer energy with less efficiency (about 40%). Cyanobacteria, red algae and cryptophyceae use accessory pigments, the phycobilins, which have an energy transfer probability of close to 100%.

9.3.2 Carotenoids

Carotenoids are yellow or orange long-chain tetraterpenes; about 500 chemically different forms have been identified in the plant kingdom with about a quarter in higher plants. They are found in photosynthetically active and senescent leaf tissues, in fruits and petals, as well as in anthers, pollen and

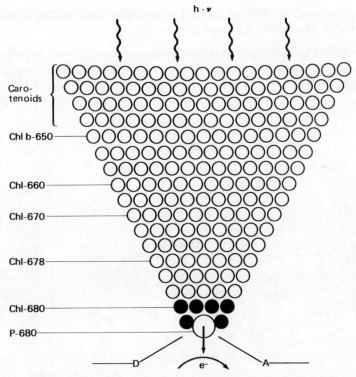

Fig. 9.10. Schematic presentation of the effective LH pigment array (energy trap) of photosystem II (from Richter, 1982, modified).

seeds. The main carotenoids in green tissues are β-carotene, lutein, violaxanthin, neoxanthin, antheraxanthin and zeaxanthin. In the photosynthetic units of the thylakoids they are probably membrane-bound and act as accessory pigments. Due to their long-chain tetraterpene structure carotenoids are very lipophilic. They are distinguished into the oxygen-free carotenes and the oxygen-containing xanthophylls (Fig. 9.11).

Carotenoids are isomeric hydrocarbons with a central chain of isoprenoid units and terminal α- or β-ionon structure. α-carotene is optically active (i.e. it turns the plane of polarized light) due to its asymmetrically substituted carbon atom in the terminal ring. In contrast, β-carotene is symmetrical with 11 conjugated double bonds in all-*trans* positions as in all naturally occurring carotenoids. β-carotene is found, for example, in carrots and is used by animals and men, which usually cannot synthesize carotenoids as essential provitamin A which is necessary for vision (see Chapter 13).

Xanthophylls contain oxygen in the form of hydroxy, epoxy, carboxy or methoxy groups. Lutein is the most common xanthophyll in leaves, and contains two hydroxy groups; spirilloxanthin, the typical carotenoid in purple bacteria has two methoxy groups (Fig. 9.11).

In addition to xanthophylls yellow leaves and petals, as well as yellow and red fruits, contain xanthophyll esters. Their absorption spectra are identical with those of the corresponding xanthophylls (Fig. 9.12).

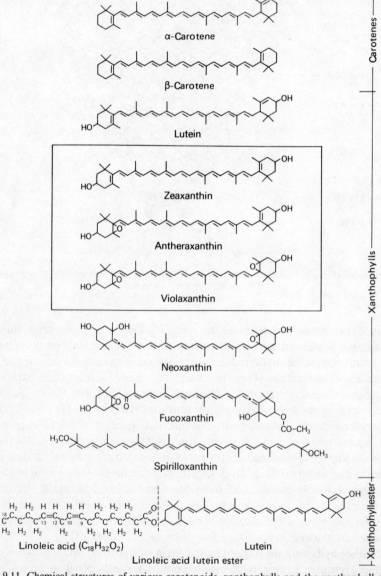

Fig. 9.11. Chemical structures of various carotenoids, xanthophylls and the xanthophyll ester luteine linoleate. The xanthophylls involved in the xanthophyll cycle are marked by box.

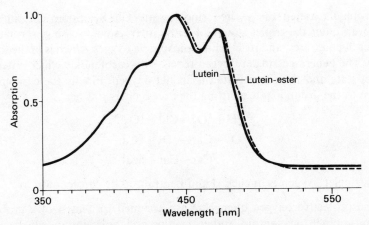

Fig. 9.12. Absorption spectra of lutein and lutein esters recorded by a fast photometer during high-pressure liquid chromatography (from Tevini *et al.*).

9.3.2.1   FUNCTION

Carotenoids absorb blue radiation and usually have a three-peaked spectrum. Lutein, for instance, absorbs at 425 nm, 445 nm and 475 nm in methanol. In photosynthesis, carotenoids serve as accessory pigments. Due to their broad absorption bands with several peaks they widen the blue spectral range available for photosynthetic energy harvesting. In higher plants and green algae the energy absorbed by carotenoids is transferred with a probability of only 40–50%. In brown algae and diatoms, however, fucoxanthin, which causes the brown color of these algae, transfers its excitation energy to the reaction centers almost without losses.

In addition to their role as accessory pigments, carotenoids protect chlorophylls against photooxidation. Mutants of *Zea mays*, for instance, which synthesize only phytoen, phytofluen and ζ-carotene contain chlorophylls in weak light. In strong light, however, the chlorophylls are destroyed in the presence of oxygen but not in the presence of nitrogen. The number of conjugated double bonds of the carotenoids present seems to be important. Mutants which synthesize only carotenoids with seven conjugated double bonds are more photosensitive than mutants capable of synthesizing lycopin, which contains 11 conjugated double bonds. Some herbicides, such as the pyridazinon norfluorazon, block the biosynthesis of carotenoids. Their phytotoxic action is based on the fact that the chlorophylls are oxidized by light in carotenoid-free tissues.

The protective function of carotenoids is due to the inactivation of singlet oxygen which is produced in light in the presence of a photosensitizer. In photosynthesis chlorophyll acts as a photosensitizer. The majority of the chlorophylls are in the excited singlet state Chl-S_1. Some chlorophylls, however, enter the triplet state Chl-T_1 by intersystem crossing, specially at higher fluence rates, and react with atmospheric oxygen which is in the triplet state. The generated singlet oxygen reacts with carotenoids which enter the triplet state and thus deactivate the singlet oxygen. Finally the carotenoids return to the ground state emitting the excess energy as heat.

$$^3Chl + {}^3O_2 \rightarrow Chl + {}^1O_2$$
$$^1O_2 + Car \rightarrow {}^3O_2 + {}^3Car$$
$$^3Car \rightarrow Car + heat$$

Thus, singlet oxygen is quenched and neutralized by carotenoids.

The protective oxygen quenchers are assumed to consist of a group of xanthophylls, zeaxanthin, antheraxanthin and violaxanthin, which can be converted into each other according to the following scheme (see Fig. 9.11):

zeaxanthin↔antheraxanthin↔violaxanthin.

This process, called xanthophyll cycle, is driven by photosynthetic electron transport and is found in higher plants, green and brown algae.

9.3.3   Phycobilins

The photosynthetic action spectra of red algae and cyanobacteria (blue–green algae), in contrast to that of green algae and higher plants, show strong activity in the green region. The absorption spectrum of the red alga *Porphyridium cruentum* has a wide band in the green range with maxima at 547 nm and 565 nm centered between the chlorophyll absorption maxima. Absorption and activity are due to specific accessory pigments with phycobilins as prosthetic groups. Like the chlorophylls, the phycobilins consist of a tetrapyrrole system which is linearly arranged and in contrast to all other accessory pigments covalently bound to an apoprotein (Fig. 9.6). *In vivo* we find several phycobiliproteins which can be distinguished into three groups: blue allophycocyanin, and blue phycocyanin, which both contain phycocyanobilin as a prosthetic group and red phycoerythrin with phycoerythrobilin as a prosthetic group. These forms are found in red algae and cyanobacteria in various molar ratios. Recently, in *Anabaena*, a phycoerythrocyanin was found that has three chromophores, one of which is phycobiliviolin.

The phycobiliproteins are arranged in structural units, the phycobilisomes, which are attached to the thylakoids of red algae and cyanobacteria. Electron micrographs of phycobilisomes isolated from *Synechocystis* (PCC 6701) show a regular substructure of several subunits which are assumed to correspond

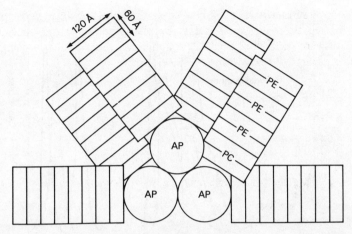

Fig. 9.13. Schematic presentation of the cross-section through a phycobilisome of cyanobacterium LPP-7409, grown in green light. PE = phycoerythrin, PC = phycocyanin, AP = allophycocyanin (from Stumpf).

with the phycobiliproteins. Figure 9.13 shows the schematic molecular arrangement of the phycoerythrin, phycocyanin and allophycocyanin in a cyanobacterium in a molar ratio of 3:1:1.

The spectral properties of the phycobilins depend on their prosthetic groups and their aggregation with the protein. Phycoerythrins absorb between 498 and 567 nm, phycocyanins between 555 and 620 nm and allophycocyanins between 618 and 671 nm (Fig. 9.14). Phycobiliproteins with trimeric substructures have absorption maxima at wavelengths longer than those with monomeric substructures.

Phycobiliproteins act as light-harvesting complexes and transfer their excitation energy to chlorophylls in the photosystems with an effectivity of about 100%. The energy transfer to photosystem II is probably more efficient than that to PSI since PSII is, as apparent from electron micrographs, directly adjacent to the phycobilisomes.

9.4 PHOTOSYNTHETIC ELECTRON TRANSPORT

Photosynthetic electron transport is started by the absorption of photons. Accessory pigments, integrated in the LHC, transfer the absorbed energy to chlorophyll–protein complexes, which in turn funnel it to the reaction centers of photosystems I and II. The current concept of two separate photosystems with two separate light reactions dates back 25 years to observations by

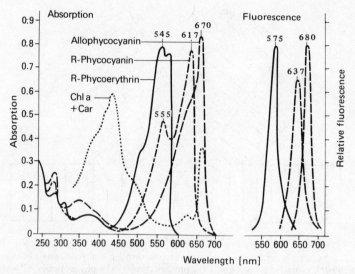

Fig. 9.14. Absorption and fluorescence emission spectra of the biliproteins in the phycobilisomes of the red alga *Porphyridium cruentum* compared with the absorption spectrum of a mixture of chlorophyll a and carotenoids (from Gantt and Lipschultz).

Emerson. He found that the quantum yield of the photosynthetic O_2 production in green algae drastically decreases beyond 680 nm. This so-called red drop does not occur when photons of 650 nm and 700 nm are irradiated simultaneously. O_2 production is even higher than the sum of the O_2 productions when 650 nm and 700 nm quanta are irradiated separately. This effect is called the 'Emerson enhancement effect'; it can only be explained by assuming the activity of two photoreactions. The Emerson effect can even be observed when the sample is irradiated alternatively with the two wavelengths at time intervals of up to several seconds. This indicates that there are light-independent steps connecting the two photoreactions.

The concept of an electron transport chain connecting two photosystems in series is found in all photosynthetic organisms which develop oxygen from water and reduce NADP (Fig. 9.15a). In contrast, green bacteria (Fig. 9.15b) and purple bacteria (Fig. 9.15c) have an anoxygenic photosynthesis with only one photosystem (see Section 9.5).

9.4.1   Electron transport and primary reactions of the photosystems

The components of the electron transport chain are often drawn in the form of a zig-zag diagram (Z scheme) indicating their redox potentials (Fig. 9.15a). The Z scheme shows the energy levels of the electron flux along the

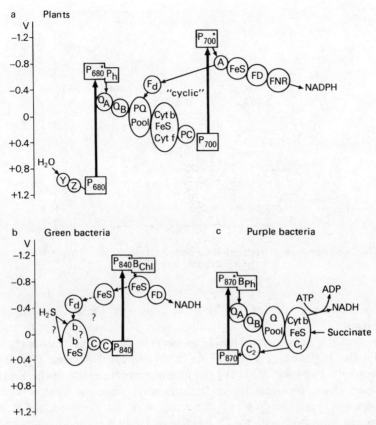

Fig. 9.15. Schematic diagrams of the photosynthetic electron transport chain in the form of a potential diagram in (a) higher plants, (b) green bacteria and (c) purple bacteria (from Wright, modified).

chlorophylls and the redox systems, from water with a low free energy (positive redox potential) to NADP with a high free energy (negative redox potential). The electron transport chain starts with the photolysis of water supplying the electrons which, via intermediates, reach the reaction center of photosystem II, P680. P680 is the electron donor for the acceptor plastoquinone (PQ) from which the electrons are transported via the redox systems cytochrome f and plastocyanin (PC) to the reaction center of photosystem I, P700. P700 is the electron donor for some iron- and sulfur-containing redox systems which funnel the electrons via some intermediates to the terminal acceptor NADP. NADP is reduced to NADPH under simultaneous uptake of protons. The potential difference between water and NADPH is thus covered in two steps driven by the absorption of radiation energy in the chlorophylls. In reality this

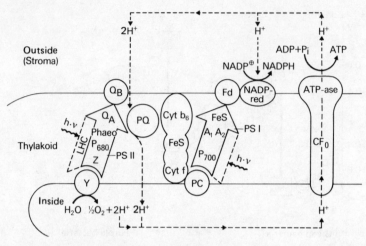

Fig. 9.16. Two-dimensional presentation of the electron transport chain of photosystems I and II in the thylakoid membrane.

very schematic model of the electron transport chain is much more complicated.

The spatial arrangement of the components in the electron transport chain within the thylakoid are shown in a hypothetical scheme (Fig. 9.16). Chlorophylls P680 and P700 are the photochemical centers of the electron transport chain. By using the excitation energy from the light-harvesting complexes they are able to supply the acceptor A with electrons, and in turn receive electrons from a donor D. P (P680 or P700) is excited by light which induces a charge separation by which P is oxidized and the acceptor is reduced. The missing electron of P is restored by a transfer from D to P. The oxidized donor D^+ is regenerated by the uptake of an electron and A^- by the release of the electron (Fig. 9.17). After this P can be excited again.

9.4.1.1   PHOTOSYSTEM II

Chlorophyll P680 is the center of photosystem II and receives electrons from water using the oxygen-evolving system. After excitation P680 functions as a primary electron donor and sends an electron to a specifically bound plastoquinone Q_A via phaeophytin, which acts as an intermediate. By this process P680 is oxidized ($P680^+$) and receives its electron back from the donor Z. Z eventually gets its electron back from water. The splitting of water is catalyzed by a manganese containing enzyme complex Y. The liberation of a molecule of oxygen from two H_2O molecules only occurs after four oxidized

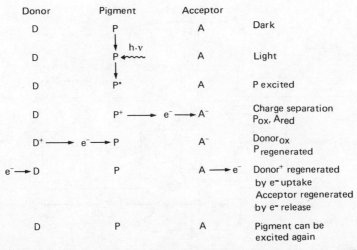

Fig. 9.17. Schematic presentation of the photosynthetic processes in the reaction center.

components Y^+ to Y^{4+} have been accumulated. Joliot demonstrated that chloroplasts develop oxygen only after a sequence of four extremely short flashes:

$$2H_2O \xrightarrow{hv} Y \xrightarrow{hv} Y^+ \xrightarrow{hv} Y^{2+} \xrightarrow{hv} Y^{3+} \xrightarrow{hv} Y^{4+} \longrightarrow O_2 + 4H^+ + Y$$

After a previous dark phase the first maximal O_2 burst is found after three flashes and then again after each four flashes (after the 7th, 11th, 15th . . . flash). It is assumed that in darkness Y^+ already exists in a stable form which is converted into Y^{2+} by the 1st flash. During oxidation protons are transported into the interior of the thylakoid (Figs 9.16 and 9.18). The redox potential of the reaction described above is $+0.82$ V. Since P680$^+$ oxidizes water its potential needs to be even higher.

The high redox potential hinders its measurement, since by accumulating P680$^+$ neighboring molecules can also be oxidized. However, by measuring difference spectra fast absorption changes can be detected. Chloroplasts are irradiated with strong short flashes of light (10 μs) and the resulting absorption changes (light minus dark) are plotted in dependence of the wavelength. A negative peak means a light-induced absorption decrease. The absorption changes have different kinetics with an increase of 20 ns and a decrease of 0.2 ms (Fig. 9.19). The maximal absorption change of P680 is found at 680 nm. Another peak at 430 nm is supposed to represent the dimeric form of the P680 chlorophyll. Dimeric chlorophyll forms ('chlorophyll special pairs') are currently considered as the reaction centers of both photosystems.

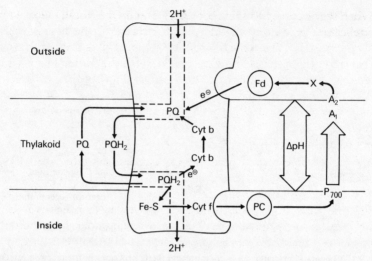

Fig. 9.18. Schematic presentation of the spatial arrangement of the electron transport chain (cyclic) involving the integral FeS–cytochrome complex pumping protons (PQ cycle) (from Barber, modified).

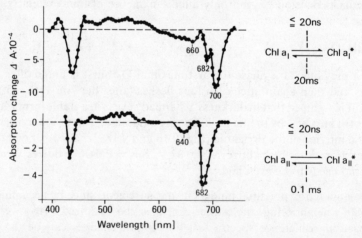

Fig. 9.19. Difference spectra of P_{700} (above) and P_{680} (below) of isolated chlorophyll protein complexes. Right: kinetics of the absorption changes (from Kok and Döring et al.).

On the acceptor side of PSII, the electron transport chain starts with two plastoquinone forms Q_A and Q_B. Q_A effectively quenches the chlorophyll fluorescence. Difference spectra and the Kautsky fluorescence induction curve allow speculation on the redox potential of these first stable electron acceptors Q_A and Q_B (see Fig. 9.9).

When the electron transport between Q_A and Q_B is totally blocked by the application of diuron we find a fast increase of the fluorescence to a maximal value. The reduction of Q (Q_A and Q_B) is accompanied by an absorption change at 320 nm attributed to the generation of a plastosemi-quinone. Therefore the term X320 is also used for the primary acceptor of PSII. Another absorption change is found at 550 nm (C550). It is due to a shift in the absorption band of a β-carotene exposed to a strong electric field.

The existence of a second quinone, Q_B (sometimes only called B), as an electron transmitter between Q_A and the following PQ pool can be deduced from the observation that the electron transfer from Q_B into the PQ pool occurs only after two flashes. Thus, Q_B is reduced to plastohydroquinone (PQH_2) in two subsequent reactions via a semiquinone by taking up two protons. It is separated from the binding site at the PSII complex as PQH_2 and diffuses into the PQ pool. The Q_B binding site then accepts a new PQ molecule from the PQ pool.

It has long been known that the effect of many agriculturally important herbicides depends on the inhibition of the photosynthetic primary reactions. Typical examples for these substances are urea compounds, such as diuron and triazine. But it was found only recently that the mechanism depends on competition between the herbicide and the acceptor quinone Q_B. When the herbicide binds to the Q_B binding site of the PSII complex it is no longer available for plastoquinone and the electron transport chain is interrupted.

Between the PQ pool and PSI there is an integral protein complex, which contains a cytochrome f in addition to a cytochrome b_6 ($=$ cyt 563) and a Fe–S protein. This complex (also called the Rieske center) has a binding site for PQH_2, probably at the Fe–S protein. During oxidation of PQH_2, protons are released and transported into the thylakoid interior. This causes an acidification of the inner compartment similar to the one produced during water splitting (Fig. 9.18). The role of cytochrome b_6 is still obscure. The electron transport continues from the Rieske center via cytochrome f to plastocyanin, a copper-containing protein, which serves as the electron donor for PSI. Cytochrome f and plastocyanin can be functionally localized by measuring fast absorption changes and by using inhibitors.

When P700 is rapidly oxidized by a flash its reduction in darkness shows several phases of different lengths indicating several pools of plastocyanin. One pool is membrane-bound while the other is freely motile on the thylakoid inner surface. The functional significance of these two pools is still obscure.

9.4.1.2 PHOTOSYSTEM I

Results obtained with absorption, difference and electron spin resonance spectroscopy are the basis of the hypothetical arrangement of the primary and secondary electron acceptors of PSI in chloroplasts shown in Fig. 9.16.

The reaction center of PSI is thought to consist of two chlorophyll molecules or a chlorophyll dimer, because of the two absorption maxima found in the difference spectrum at 682 nm and 700 nm. After excitation the electrons are probably accepted by a special chlorophyll a form A_1 before the primary recently identified electron acceptor vitamin K_1 (A_1), is reduced. Membrane-bound iron–sulfur complexes (Fe–S_A and Fe–S_B) are the secondary electron acceptors. The acceptor(s) have to have an even more negative redox potential than ferredoxin (Fd, $E° = -420$ mV) which is thought to be the following electron acceptor. In contrast to ferredoxin which is loosely bound to the thylakoid surface the NADP reductase is a membrane-bound flavoprotein ($E° = -380$ mV) which catalyzes the electron transport from ferredoxin to NADP. Protons are taken up from the stroma in order to reduce NADP which concludes the noncyclic (linear) electron transport from water to NADP via photosystems II and I.

In addition to linear electron transport, photosystem I can also mediate cyclic electron transport. Starting from the reaction center P700 an electron follows a path via the Fe–S proteins and ferredoxin back to the plastoquinone pool rather than to NADP. By accepting a second electron from cytochrome b_6 and protons from the stroma, PQ is reduced to PQH_2. From here the electrons follow the linear electron transport back to P700, whilst cytochrome b_6 serves as an oxidant for PQH_2 (Fig. 9.18). Thus the cyclic electron transport is closed without losing any electrons. During this cycle no NADP is reduced but protons are transported vectorially through the thylakoid membrane which also generates a proton gradient between thylakoid inside and outside. The electrochemical potential difference of this proton gradient is utilized to produce ATP.

9.4.2 Artificial electron transport and inhibitors

By using inhibitors of the electron transport chain it is possible to isolate functional parts of the chain and to analyze partial reactions spectroscopically. Furthermore, it is possible to follow the electron flux over some redox systems by using artificial electron donors and acceptors. Hill (in 1939) was the first to successfully reduce an artificial electron acceptor by a green cell homogenate. He first used iron oxalate and later potassium ferricyanide (Fe^{3+}) which is reduced to Fe^{2+} by electron uptake. Today intensely colored dyes are used for the Hill reaction, such as dichlorphenolindophenole (DCPIP), which is deep blue in the oxidized state and colorless in the reduced state. By irradiation of a dilute chloroplast suspension the bleaching can be followed by the naked eye.

In the meantime the number of artificial electron donors and acceptors has increased enormously. In addition there are inhibitors, especially herbicides, which block electron transport in well-defined positions. A selection of inhibitors and Hill reagents and their action sites are summarized in Fig. 9.20.

Pseudocyclic electron transport is a variant of cyclic electron transport, and possibly even occurs *in vivo*. Electrons are transported along the linear chain but are not used to reduce NADP but return to molecular oxygen. Ferredoxin is an autoxidable intermediate. From a physiological point of view electron transport back to O_2 is only useful when enough NADP is in the reduced form and cannot be reoxidized in the Calvin cycle due to a lack of CO_2. This situation can occur at high energy fluence rates and limited gas exchange. In isolated chloroplasts electron transfer to O_2 has been experimentally demonstrated with methylviologen (MV) as the electron acceptor. This so-called Mehler reaction uses two molecules of O_2 per one photosynthetically produced molecule O_2. Thus the Mehler reaction is a mirror-image of the photosynthetic production of O_2 and is a Hill reaction including the whole electron transport chain. In addition to H_2O_2, a number of reactive, toxic oxygen radicals are produced in the Mehler reaction, catalyzed by methylviologen, which oxidize various membrane components such as pigments and lipids. Therefore methylviologen (= paraquat) is used in agriculture as a herbicide.

9.4.3   The cooperation of the photosystems

The absorbed photoenergy is best utilized when the photosystems are optimally synchronized in noncyclic electron transport. Since some of the thylakoid proteins have a rapid turnover the thylakoids can quickly adapt to the changing biochemical conditions. In addition the number and composition of photosystems in the grana and stroma thylakoids is extremely variable, depending on the light conditions.

Chloroplasts possess a self-regulating system which optimally distributes the excitation energy to the two photosystems. It consists of a kinase, the activation level of which depends on the redox state of the PQ pool, and which phosphorylates proteins located on the surface of LHCII (Fig. 9.21). Protein phosphorylation induces higher charges in the grana stack, and the grana thylakoids move further apart. As a consequence the LHC proteins of photosystem II become more mobile and can diffuse laterally toward the

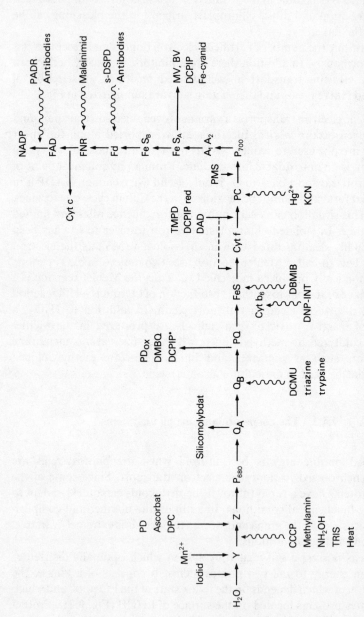

Fig. 9.20. Inhibitors (wavy lines) and artificial electron donors and acceptors (straight arrows) in the photosynthetic electron transport (from Richter, modified and extended). BV = benzylviologen, CCCP = carbonyl cyanide chlorophenylhydrazone, DAD = diaminodurole, DBMIB = dibromomethyl isopropyl benzoquinone, DCPIP = dichlorophenolindophenole, DCMU = diuron = dichlorophenyldimethylurea, DPC = diphenylcarbazid, s-DSPD = sulfodisalicylidin propandiamine, DNP-INT = dinitrophenoxyl nitrobenzene, DMBQ = dimethylbenzoquinone, MV = methylviologen, PADR = phosphoadenosine-P-P-ribose, PD = p-phenylendiamine, PMS = phenazinemethosulfate, TMPD = tetramethyl-p-phenylenediamine.

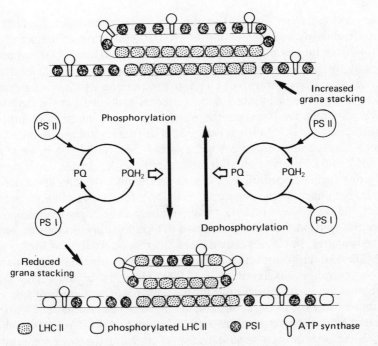

Fig. 9.21. Model showing the control of the excitation energy distribution in photosynthesis by PQH_2 induced phosphorylation of LHC II which causes a reduced grana stacking (from Barber).

photosystem I particles in the stroma region. This process requires a certain fluidity of the lipid matrix which we have already mentioned in connection with ATPase mobility. When the LHCII proteins move closer to PSI, which predominates in the stroma region, there is a better light energy distribution between PSII and PSI, and PSII escapes photooxidation by excessive light energy. In lower fluence rates PQ is basically in the oxidized state, and the proteins are dephosphorylated, causing increased grana stacking. In accordance with these observations we find more grana stacks in shade plants than in sun plants. Thus chloroplasts possess an effective control system to adapt to the light conditions using the redox state of the PQ pool.

9.4.4 Electron transport, pH gradient and photophosphorylation

Photosynthetic electron transport is a vectorial process, i.e. electrons are transported in one direction through the thylakoid membrane (Fig. 9.16). Coupled to the electron transport from water to NADP there is a transport of protons. The photolysis of water at the beginning of the electron transport chain results in a release of two electrons and two protons, which increase the

GP–F

proton concentration inside the thylakoid. Noncyclic (linear) electron transport ends with the reduction of NADP, consisting of uptake of the electrons from the chain and protons from the stroma. Thus NADP reduction decreases the proton concentration in the stroma. In addition the redox system in the electron transport chain, plastoquinone, operates as a proton carrier through the membrane, since it accepts protons from the thylakoid outside and releases them on the inside. The exact number of protons translocated in the PQ shuttle per electron is still unclear. Normally we would expect a ratio of one proton per electron but the cytochrome b_6 cycle (see Fig. 9.18) could transport an additional proton under certain experimental conditions. Including proton release during water photolysis we find a ratio of $4H^+/2e^-$ or under special conditions $5H^+/2e^-$. In bacteria only one proton seems to be transported into the chromatophore vesicles per electron.

Electron transport in light generates a pH gradient in chloroplasts and in chromatophores. In chloroplasts the pH difference between the stroma and the thylakoid interior can amount to up to 4 pH units, which corresponds to a H^+ concentration difference of $1:10,000$. Since on a short-term basis membranes are fairly impermeable for protons this proton gradient can only equilibrate using specific structures. In the thylakoid membrane, but spatially separated from the electron transport chain, we find ATP synthases which allow a vectorial proton reverse transport and use the stored energy to produce ATP. In chloroplasts the stoichiometry has been determined experimentally as one ATP per three protons; bacteria seem to need only two protons.

9.4.4.1   MECHANISM OF ATP PRODUCTION

According to Mitchell's and Jagendorf's chemiosmotic hypothesis ATP synthesis is driven by an electrochemical gradient (see Section 6.3.5) consisting of two components: a pH gradient, generated by the vectorial proton transport and the electrical membrane potential generated by electron transport from the thylakoid inside (positive) to the outside (negative). The energy stored in the electrochemical gradient is used during the proton reverse transport to synthesize ATP from $ADP + P_i$.

The responsible enzyme, the ATP synthase, consists of subunits, CF_0 and CF_1. The coupling factor CF_1, composed of five subunits, is elevated over the surface and catalyzes the ATP synthesis. CF_0 spans the membrane and seems to be responsible for channeling the protons through the membrane. ATP synthesis starts with the binding of phosphate to CF_1 which reacts with two protons transported along the CF_0 channel. A reactive metaphosphate (PO^{3-}) is formed, which reacts with ADP to yield ATP (Fig. 9.22a).

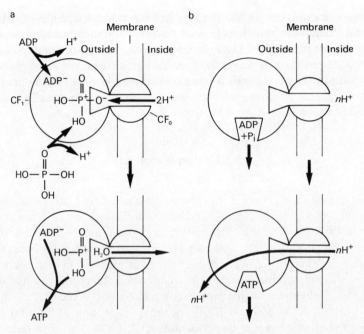

Fig. 9.22. ATP formation by proton translocation at the CF_1/CF_0-complex (from McCarty). (a) Direct mechanism from Mitchell. (b) Indirect mechanism by conformational changes of the coupling factor (after Boyer, Harris, Slater from Richter, modified).

According to an alternative hypothesis ATP synthesis takes place spontaneously at the active site of the ATP synthase from ADP and phosphate without additional free energy. Protons play only an indirect role, and cause a conformational change in the coupling factor which leads to the release of the produced ATP (Fig. 9.22). The vectorial electron transport and the proton translocation generating a membrane potential are the strongest support for the chemiosmotic hypothesis. In addition, ATP synthesis can be induced even in darkness when an artificial pH gradient or an artificial electric field are generated across the thylakoid membrane. These experimental results, however, do not exclude the conformational change hypothesis. The results found in *Halobacterium*, which is a model system for light-mediated proton translocations (see Chapter 8), also support the chemiosmotic hypothesis.

9.5 BACTERIAL PHOTOSYNTHESIS

In contrast to photosynthesis in cyanobacteria and photosynthetic eukaryotes, photosynthesis in purple bacteria and green bacteria is driven by only one photoreaction during which no oxygen is produced. The electron donors are sulfur compounds such as H_2S or organic substances taken up from the environment, rather than H_2O. Cyanobacteria, which are regarded as true bacteria due to their prokaryotic organization, have an oxygenic photosynthesis with two light reactions which closely resemble that of higher plants. As in higher plants, in photosynthetic bacteria an electrochemical gradient is generated which is used for ATP synthesis.

9.5.1   Purple bacteria

Purple bacteria are classified into the sulfur-containing Chromatiaceae and the sulfur-free Rhodospirillaceae. The latter use alcohols and carbonic acids as electron donors while the Chromatiaceae usually use H_2S. In all purple bacteria the photosynthetic apparatus is localized in chromatophores which resemble thylakoids (Fig. 9.23). These vesicles originate from invaginations of the cytoplasmic membrane and are interconnected. According to a hypothetical model, the light-harvesting protein complex consists of the bacteriochlorophyll proteins B800–850 and B870 (maximal absorption at 870 nm) which enclose the reaction center bacteriochlorophyll P870. This arrangement of antenna pigments allows a directed exciton transport similar to the one found

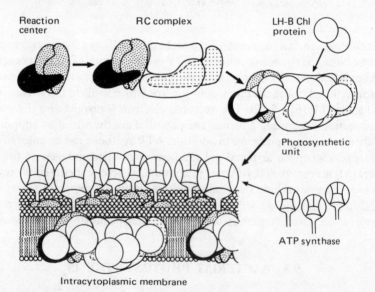

Fig. 9.23. Intracytoplasmic membrane (photosynthetic part) of photosynthetic purple bacteria and the individual particles shown separately (from left to right). LH-B-Chl = light-harvesting bacteriochlorophyll protein (from Sauer).

in higher plants. Several cytochromes also belong to the reaction center complex (Fig. 9.23).

After excitation the electrons are transported from P870, which is assumed to operate as a dimer, to the acceptor X, presumably a bacteriophaeophytin (in analogy to phaeophytin in PSII of higher plants). The electrons follow a cycle via the secondary acceptors Q_A and Q_B, the quinone pool and several cytochromes and back to P870. *Chromatium* seems to use naphthoquinone as Q_A while other purple bacteria use ubiquinone. In analogy to the Q cycle of higher plants, protons are transported through the membrane into the chromatophore interior. In purple bacteria the pH difference between cytoplasm and the chromatophore inner space is only 2 pH units in contrast to the 4 pH units found in thylakoids of higher plants. The membrane potential of about 200 mV is the main component of the electrochemical gradient necessary for the ATP synthesis.

NAD is reduced by an inverted electron transport driven by ATP; the electrons are abstracted from the ubiquinone pool and the electron deficit is compensated by external electron and hydrogen donors such as fumarate and hydrogen (Fig. 9.15c).

9.5.2 Green bacteria

The green bacteria contain chlorosomes as photosynthetic organelles. These are vesicles closely connected with the cytoplasmic membrane by a crystalline basal plate. Figure 9.24 shows a chlorosome model which has been developed on the basis of electron microscopic studies in *Chlorobium limicola f. thiosulphatophilum*. The cytoplasma membrane is thought to contain the non

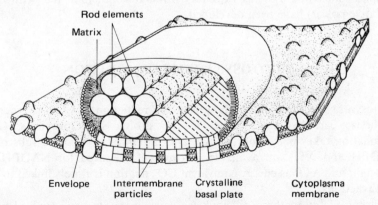

Fig. 9.24. Model of a chlorosome with the associated plasma membrane of the green bacterium *Chlorobium limicola* (from Staehelin *et al.*, 1980).

crystalline light harvesting complex and the crystalline basal plate the bacteriochlorophyll a–protein complex. Inside the chlorosome, which is surrounded by a single layer of galactolipids, we find rod-like elements which are thought to consist of bacteriochlorophyll c–protein complexes.

Bacteriochlorophyll P840 is the reaction center pigment which passes electrons to a primary acceptor with a very negative potential. In contrast to purple bacteria, this acceptor can pass electrons to NAD via ferredoxin (Fig. 9.15b). The electron deficit of P840 is probably compensated by a chain of cytochromes which eventually receive the electrons from H_2S or other donors. In analogy to oxygen production in green plants, molecular sulfur is produced and deposited outside the cell. In addition to the noncyclic electron transport, Chlorobiaceae seem to possess a cyclic one.

Thus there is a close relationship between linear electron transport and direct NAD reduction in green bacteria and the photosynthetic process in cyanobacteria and higher plants. Furthermore, the chlorosomes of the green bacteria structurally resemble the cyanobacterial phycobilisomes. Therefore green bacteria could be an evolutionary link between purple bacteria and cyanobacteria. On the other hand, the electron acceptor side (Q) of the photosystem closely resembles that of PSII in higher plants (PQ) while the donor side in bacterial photosynthesis resembles that of photosystem I in higher plants because of the cytochrome–Fe–S complex. It is possible that both photosystems II and I have developed from a common bacterial photosystem. The evolution of photosynthesis in higher plants is therefore not characterized by the development of two photosystems but by the 'invention' of water splitting activity linked with photosystem II. The energy difference between the donor and the acceptor is the same in all photosystems, but they operate on different levels. Photosystem II starts from a very low level and photolysis of water could only be performed after a chlorophyll–protein complex and redox systems with very low redox potential (i.e. with high oxidative power) have been developed.

9.6   PHOTOSYNTHETIC CO_2 FIXATION

Photosynthetic CO_2 fixation, as well as the reduction of CO_2 in order to synthesize carbohydrates, occurs in the stroma of chloroplasts. CO_2 assimilation takes place both in light and in darkness when the substrates NADPH and ATP are available. Because of the need for NADPH as reductant and ATP as energy equivalent CO_2 fixation is closely linked to the light reactions.

During evolution three different ecological variants have evolved with different CO_2 incorporation mechanisms: C_3, C_4 and CAM plants. In C_3

plants the first detectable CO_2 fixation product is a C_3 compound, phosphoglycerin acid, in C_4 plants we find C_4 compounds such as malate or oxalacetate. In CAM plants, most of which are Crassulaceae, CO_2 is incorporated into C_4 acids during the night; these are used as a CO_2 reservoir for CO_2 fixation during daytime, which follows the same pattern as in C_3 plants.

9.6.1 CO_2 fixation and reduction in C_3 plants

In C_3 plants, such as spinach, barley and wheat, CO_2 fixation occurs in the reductive pentosephosphate cycle (also called the Calvin cycle after its discoverer). CO_2 is incorporated into a phosphorylated pentose, ribulose-1,5-bisphosphate (RubP). The resulting unstable C_6 compound splits into two C_3 compounds. The first detectable product is phosphoglycerate (PGA), which is reduced to the corresponding triose. The key enzymes are RubP carboxylase, which catalyzes the CO_2 incorporation and glyceraldehyde phosphate dehydrogenase, which reduces PGA by means of NADPH after activation with ATP (Fig. 9.25). The reaction results in two molecules of triose phosphate which are eventually converted by further reactions into fructose phosphate (FruP). Regeneration of the CO_2 acceptor occurs in a complicated cycle involving C_2 to C_7 intermediates, finally resulting in the production of ribulose monophosphate, which is activated by ATP to RubP.

For the reduction of each molecule of CO_2 three molecules of ATP and two molecules of NADPH are needed. During the day starch is synthesized from FruP via glucose phosphate and stored in the chloroplasts. At night it is used as transitory starch for the catabolic metabolism.

The Calvin cycle is controlled by many factors. The enzymatic reactions are affected by the concentration of the substrates and their products. In addition the light reactions change pH and ion concentrations. Changes in the stroma pH, from 7 in darkness to 8–8.5 in light, induce several enzymes of the Calvin cycle to operate optimally, which they do not do at pH 7 in darkness. Furthermore, a number of enzymes in the Calvin cycle are directly activated by light (see Section 9.6.5). A decrease in the CO_2 or an increase in the O_2 partial pressure induces the RubP carboxylase to operate predominantly as an oxygenase, which increases the photorespiration rate (see Section 9.6.4).

9.6.2 CO_2 fixation and reduction in C_4 plants

C_4 plants, such as corn, sugar cane and millet, fixate CO_2 primarily by incorporating it into phosphoenolpyruvate (PEP). First stable intermediates

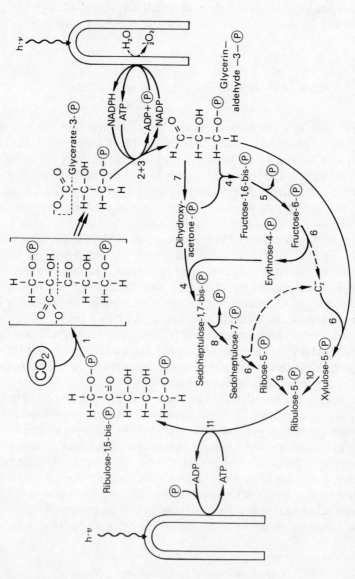

Fig. 9.25. Calvin cycle (reductive pentose phosphate cycle). 1: Ribulose bisphosphate carboxylase, 2 and 3: phosphoglycerate kinase and glycerinaldehyd phosphate dehydrogenase, 4: aldolase, 5, 8: phosphatases, 6: transketolase, 7: triosephosphate isomerase, 9, 10: pentosephosphate isomerases, 11: ribulosephosphate kinase (after Bassham, from Mohr and Schopfer, 1978, modified).

are C_4 acids and not PGA. After decarboxylation the liberated CO_2 is secondarily fixed, as it is in C_3 plants, and incorporated into carbohydrates. C_4 plants differ from C_3 plants in their leaf anatomy with a pronounced chloroplast dimorphism and functionally by a more efficient CO_2 binding mechanism which improves the apparent photosynthesis. At light saturation, which is reached at very high fluence rates, photosynthetic net production by far exceeds that of C_3 plants (Table 9.2).

TABLE 9.2   THE PHYSIOLOGICAL CHARACTERISTICS OF CO_2 FIXATION OF C_3, C_4 AND CAM PLANTS SHOWN IN COMPARISON (AFTER OSMOND *et al.*)

Feature	C_3	C_4	CAM
Optimal temperature [°C]	15–25	30–45	*ca.* 35
Light saturation of CO_2 fixation [μE m^{-2} s^{-1}]	*ca.* 1000	*ca.* 2000	*ca.* <1000
CO_2 compensation - point [μl l^{-1}]	*ca.* 50	*ca.* 5	*ca.* 5–50
Water requirement for 1 g dry weight [ml]	450–950	230–250	50–55
Growth rate [g DW dm^{-2} day^{-1}]	0.5–2	3–5	0.01–0.02
Maximal CO_2 fixation [mg CO_2 dm^{-2} h^{-1}]	15–35	40–80	0.5–0.7
Photorespiration	+	−	+
CO_2 fixating enzyme	RubP-C	PEP-C	PEP-C (night) RubP-C (day)
Chlorophyll *a/b* ratio	2.8±0.4	3.9±0.6	2.5–3.0
ATP needed	3	5	4
NADPH needed	2	3	3

Leaves of C_4 plants have a ring of bundle sheath cells around the vascular elements which, in contrast to the corresponding cells in a C_3 leaf, carry chloroplasts. The space between the two epidermal layers is filled with a loose mesophyll usually forming one layer. In the mesophyll cells, PEP carboxylase catalyzes the binding of CO_2 to phosphoenolpyruvate (PEP). The resulting oxaloacetate (OAA) is transported into the usually small and starch-free mesophyll chloroplasts where malate dehydrogenase reduces OAA to malate using NADPH. Subsequently malate is transported into the bigger, grana-free but starch-containing, bundle sheath chloroplasts and split into CO_2 and pyruvate by the malate enzyme, generating NADPH.

The liberated CO_2 is secondarily fixed in the Calvin cycle and incorporated into carbohydrates. Pyruvate dikinase regenerates the primary CO_2 acceptor PEP using ATP (Fig. 9.26).

Despite this complicated CO_2 assimilation mechanism, C_4 plants have a number of advantages over C_3 plants:

1.   PEP carboxylase has a much higher CO_2 affinity than RubP carboxylase

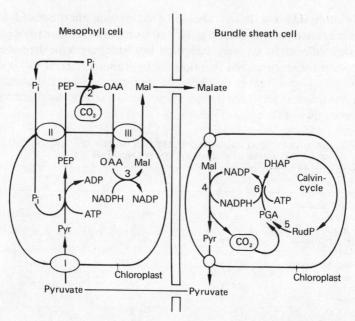

Fig. 9.26. CO_2 assimilation in C_4 plants. 1: Pyruvate dikinase, 2: PEP-carboxylase, 3: NADP-malate dehydrogenase, 4: NADP-malate enzyme, 5: RubP-carboxylase, 6: phosphoglycerate kinase, glycerinaldehyde phosphate dehydrogenase and triosephosphate isomerase; I–III carrier systems (from Edwards and Huber, 1981).

so that CO_2 can be efficiently bound even at much lower CO_2 partial pressures. Therefore the stomata need not be opened as wide as in C_3 plants, which results in a lower water loss by transpiration. Despite the less opened stomata the CO_2 turnover, and thus the photosynthetic activity, is much higher in C_4 plants than in C_3 plants at comparable fluence rates. The CO_2 compensation point is reached in C_4 plants at one-tenth of the CO_2 concentration found in C_3 plants.

2. The CO_2 concentration in the bundle sheath chloroplasts favors the carboxylation activity of the RubP carboxylase and suppresses its oxygenase activity. In addition, these chloroplasts generate less O_2 because of their low PSII activity and hardly show photorespiration. Furthermore, the PEP carboxylase would trap the CO_2 before it leaves the mesophyll cells.

For the reasons described above, C_4 plants are able to live in extremely dry, hot and sun-exposed habitats in which C_3 plants would die due to high water losses. However, C_4 plants can also grow under normal climatic conditions, as shown by agricultural corn production.

In C_4 plants CO_2 assimilation needs five molecules of ATP and two molecules of NADPH per molecule of CO_2. A more efficient use of CO_2 at low concentrations is traded for a higher ATP consumption. However, the bundle sheath chloroplasts do not seem to have a problem supplying the necessary ATP because of the increased PSI activity (hardly any grana) with cyclic photophosphorylation. Due to suppressed photorespiration, C_4 plants lose less organic material in light than C_3 plants and allow a more effective utilization of the assimilated substance at a lower water consumption. Corn and sugar cane are agriculturally important, and many weeds, such as some *Chenopodium* species, are among the C_4 plants.

9.6.3   CAM plants

The third ecological variety, CAM plants, is found in many Crassulaceae such as *Sedum* and *Kalanchoe*. Primary CO_2 fixation is followed by a diurnal acid metabolism. Therefore these plants are called crassulaceae-acid-metabolism (CAM) plants. CAM plants use an initial CO_2 fixation followed by the final incorporation using the Calvin cycle, as do C_4 plants. In contrast to C_4 plants, primary CO_2 fixation occurs at night when the stomata are wide open and allow an effective gas exchange. During daytime the stomata are almost closed so that CAM plants are not threatened with water loss by transpiration.

The CO_2 necessary for photosynthesis is stored in the endogenous CO_2 donor, malate. At night malate is formed in the cytoplasm by malate dehydrogenase, from oxaloacetate (OAA), which is produced from PEP by the activity of the PEP carboxylase as in C_4 plants. During daytime the malate accumulated in the vacuole serves as a CO_2 donor for CO_2 fixation in the Calvin cycle. PEP is produced from the transitory starch by glycolytic degradation in the chloroplasts during the night.

The ecological advantage of CAM plants is the reduced transpiration during daytime so that CAM plants are well adapted to dry and hot habitats. Due to the lack of a 'real' photosynthesis during daytime the growth rate is far lower than in all other plants (Table 9.2). With the exception of pineapples, there are hardly any agriculturally important CAM plants.

9.6.4   Photorespiration

During photorespiration O_2 is consumed and CO_2 generated in light. Photorespiration is independent of mitochondrial respiration but occurs in chloroplasts and peroxisomes and only a few reactions occur in mitochondria (Fig. 9.27).

Photosynthetic O_2 production increases the O_2 partial pressure in the

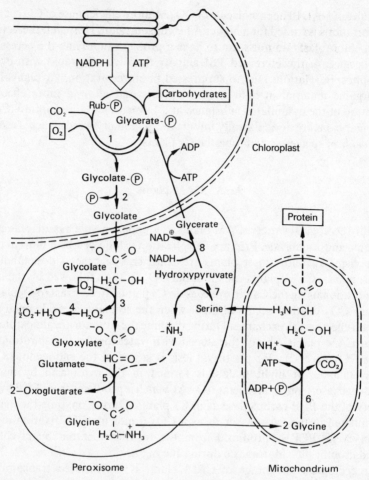

Fig. 9.27. Photorespiratory glycolate cycle. 1: Ribulose bisphosphate carboxylase, 2: phosphogly-colate phosphatase, 3: glycolate oxidase (FMN as prosthetic group), 4: catalase, 5: glutamate glyoxylate aminotransferase, 6: desamination and decarboxylation, 7: serine glyoxylate aminotransferase, 8: hydroxypyruvate reductase (after Tolbert, from Mohr and Schopfer, 1978, modified).

chloroplast. The generated O_2 competes with CO_2 for the substrate RubP, which is either split oxidatively or carboxylated in the Calvin cycle. Whether the oxygenase or carboxylase reaction predominates depends, among other factors, on the O_2 and CO_2 concentrations, as has been shown in the green alga, *Chlamydomonas reinhardtii*. An increase in O_2 concentration increases the oxygenase reaction. Both reactions are catalyzed by the same protein, fraction I protein, at the same site.

RubP is split, under O_2 consumption, into the C_2 compound phosphogly-colate and the C_3 compound phosphoglycerate by the activity of an oxygenase. After dephosphorylation glycolate can be oxidized to glyoxylate either directly in the chloroplast or in the peroxisomes. This reaction also produces the aggressive H_2O_2 which is converted into water and O_2 by catalase, the key enzyme of the peroxisomes. The glyoxylate glycine transaminase produces the amino acid glycine. Two molecules of glycine can be condensed generating CO_2 to form serine, either in the mitochondria or in the chloroplasts (Fig. 9.27). Since these reactions are dominated by C_2 compounds, photorespiration is also called the C_2 cycle.

The significance of the C_2 cycle is not yet completely clear, since some plants have a photorespiration cycle while others do not. In addition to the production of amino acids it may have a role in photoinhibition (see Section 9.7.1.1). According to this concept the excessive oxygen produced at extremely high fluence rates would be deactivated by the C_2 cycle and the pigments and proteins, especially in PSII, would be protected from oxidation. The disadvantage of photorespiration is the loss of fixed carbon which can amount to 25% of the primarily fixated CO_2. Experiments to genetically eliminate or reduce photorespiration have not yet been successful.

9.6.5   Light activation of photosynthetic enzymes

Up to now we have considered light only as an energy source for the photochemical primary processes which cause the production of NADPH and ATP starting immediately after irradiation commences. In contrast to the primary reactions, the biochemical secondary reactions in the Calvin cycle reach their optimal activity only after a lag phase of a few minutes. The delay in CO_2 fixation is due to light activation of enzymes. Two mechanisms have been found:

1. Enzyme activation occurs by means of mediators which are produced in the photoreactions (Fig. 9.28a). One of the examples for this process is the activation of triose phosphate dehydrogenase by reduced thioredoxin. Reduced ferredoxin, generated by photosynthetic electron transport, reduces thioredoxin by means of thioredoxin reductase (Fig. 9.28b). Thioredoxin in turn reduces the disulfide groups of an enzyme, such as fructose diphosphatase or NADP glycerinaldehyde dehydrogenase, causing activation by a conformational change. In darkness these light-regulated enzymes are deactivated by oxidation by, for instance, oxidized gluthathione, which is found in the chloroplast (Fig. 9.28c).

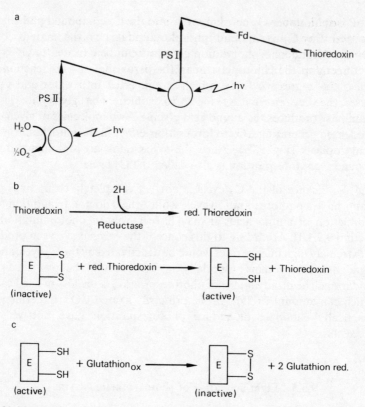

Fig. 9.28. Model of light activation and dark activation of chloroplast enzymes. Fd = ferredoxin, E = enzyme, SH = sulfhydryl group, (from Buchanan, Edwards and Walker).

2. Enzymes are activated by a change in pH or ion concentration in the stroma. Examples for this mechanism are RubP and PEP carboxylase which operate optimally at a pH of 8 and an increased Mg^+ concentration. These conditions are produced in light by vectorial proton transport into the thylakoid interior and antiparallel Mg^+ transport into the stroma. In addition the activity is affected by a change in CO_2/O_2 partial pressures (see Section 9.6.2). Furthermore, RubP carboxylase is thought to be induced by visible light and UV-A radiation according to the above-described activation by mediators.

9.7   PHYSIOLOGY OF PHOTOSYNTHESIS

Photosynthesis depends on a number of external factors, among which especially light, temperature and water supply undergo large diurnal changes.

During evolution, terrestrial plants have adapted to habitats with extreme environments by changes in the morphology and anatomy of leaves as well as the structural organization of the photosynthetic apparatus. C_4 and CAM plants have adapted to extremely dry and hot habitats with a high fluence rate, and shade plants have adapted to poor light conditions in the shade of trees.

Measurable parameters for photosynthetic production are CO_2 consumption in the Calvin cycle, O_2 production by water photolysis and dry weight production. Measuring gases, however, one has to be aware of the fact that their changes are affected both by photosynthesis and respiration. Therefore, in an intact leaf we always measure the net O_2 production (apparent photosynthesis) which is the difference between the real photosynthetic O_2 production (gross photosynthesis) and the simultaneous O_2 consumption in respiration, which includes both mitochondrial dark respiration and photorespiration (see Section 9.6.4). In an irradiated leaf photorespiration is basically responsible for oxygen consumption, since the mitochondrial respiration is inhibited by light (Kok effect).

Figure 9.29 shows schematically the relationship between apparent photosynthesis and the affecting parameters at the beginning of irradiation in dependence of the fluence rate. The light curve crosses the abscissa at a certain

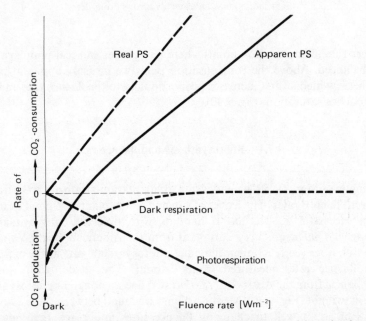

Fig. 9.29. Schematic fluence rate–response curve (light saturation curve) of the real and apparent photosynthesis and of photorespiration and dark respiration near the zero point (after Mohr and Schopfer, 1978, modified).

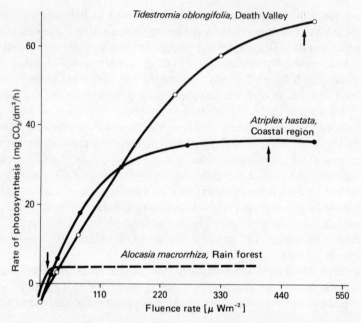

Fig. 9.30. Effect of light on the photosynthesis rate of three plant species from different habitats. Light saturation is indicated by arrows (from Berry).

fluence rate (compensation point) where respiration and real photosynthesis are balanced. Above the compensation point we measure apparent photosynthesis, which at first increases proportionally to the fluence rate and then approaches saturation (Fig. 9.30).

9.7.1   Photosynthesis and fluence rate

During evolution plants have developed adaptation strategies to extremely different fluence rates. The leaves of sun and shade plants are good examples for a successful adaptation to habitats with different light availability. *Tidestromia oblongifolia* is a sun plant from the American Death Valley; its leaves do not even reach their maximal photosynthetic rate at the extremely high fluence rates measured in its habitat. The shade plant *Alocasia macrorrhiza* from the Australian rainforest shows an apparent photosynthesis even in twilight (Fig. 9.30). Most C_3 plants are found between these extremes; they can adapt well to changing fluence rates and reach their maximal photosynthesis in normal daylight. Shade plants have a far lower compensation point than sun plants, which is due not to a lower respiration but to the fact that their photosynthetic apparatus is tailored for more efficient light utilization. Thus, shade plants use light more efficiently at low fluence rates than do sun plants. With increasing fluence rates light saturation is reached much earlier in shade plants than in sun plants: shade plants reach their maximal apparent photosynthesis at about 5% of the fluence rate of normal daylight. In contrast, sun plants need full daylight of about 360 W m^{-2} (about 2000 μE m^{-2} s^{-1}) for maximal photosynthesis, which is about 5–10 times higher than that of shade plants. Full sunlight kills many shade plants, while sun plants can survive in the shade, as shown in identical genotypes of the sun plant *Atriplex* grown under sun and shade conditions.

RubP carboxylase is thought to be the limiting enzyme for photosynthesis. In fact, shade plants always contain less of this CO_2 fixation enzyme than sun plants. Accordingly a high carboxylation rate is found in sun plants. It is obvious that the electron transport capacity needs to be high to meet the requirements for reduction and energy equivalents for CO_2 fixation. The higher photosynthetic capacity of sun plants seems to be due to a higher number of electron transport chains per chloroplast indicated by a higher ratio of cytochrome f to chlorophyll.

Thus the advantage of shade plants lies not in a high photosynthetic rate but in a more effective light exploitation. Leaves of shade plants usually have larger leaf areas and are thinner than those of sun plants. Similar differences are found between the shade and sun leaves of trees. And it is not surprising that shade plants in the tropical rainforest, such as *Alocasia macrorrhiza*, have an almost equally high chlorophyll content per leaf area unit as do sun plants, in order to effectively exploit the low fluence rates. The chloroplasts of shade plants usually contain huge grana stacks oriented in all directions in order to also utilize scattered light effectively. The high content of accessory chlorophyll *b* indicates that the LHC is larger in shade plants.

9.7.1.1 PHOTOINHIBITION

High fluence rates above the light saturation of photosynthesis cause an inhibition of photosynthesis and a bleaching of the chlorophylls in many plants. The photoinhibition is due to a biochemical change of the 32 kD protein (herbicide binding protein, HBP-32) involved in the electron transport chain of photosystem II. In contrast to the action of herbicides, such as diuron, which cause a conformational change of HBP-32, light seems to operate via a formation of peroxide which causes a biochemical change in HBP-32. Oxygen radicals and H_2O_2 are formed when electrons cannot be taken up by the electron transport chain but are returned to molecular oxygen by the semiquinone Q_B bound to HBP-32. The resulting oxygen radicals change or destroy HBP-32 and oxidize and bleach chlorophylls. However, light also stimulates the repair of the destroyed proteins by an enhanced turnover. Therefore photoinhibition occurs only when HBP-32 destruction exceeds biosynthetic renewal.

9.7.2   Radiation quality

The absorption properties of the chlorophylls involved determine the spectral sensitivity of photosynthesis (see Section 9.3). The physiological employment of radiation can be determined by measuring action spectra or spectra of the quantum yield.

In order to determine a photosynthetic action spectrum under steady-state conditions we first have to measure photosynthesis (for instance in terms of CO_2 fixation) as a function of the fluence rate at the relevant wavelengths (Fig. 9.31). In a bean leaf, for example, we need only a quarter of the fluence rate at 662.5 nm as compared to 700 nm for the same CO_2 consumption. Though the chlorophyll–protein complex has a relatively high absorption at

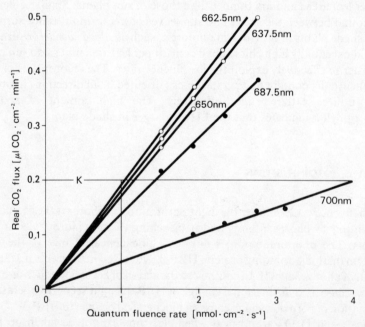

Fig. 9.31. Fluence rate–response curves of CO_2 exchange in bean leaves (from Mohr and Schopfer, 1978, modified). K = CO_2 flux of 0.2 $\mu l \, cm^{-2} \, min^{-1}$, for which about four times as many photons are needed at 700 nm than at 662.5 nm.

700 nm, this wavelength is photosynthetically rather inefficient, a phenomenon we have already discussed (see Section 9.3, red drop and Emerson
enhancement effect). When we convert the energy fluence rate into a quantum
fluence rate we can plot CO_2 consumption per mol quanta, which yields a
spectrum of the quantum efficiency (Fig. 9.32).

Comparison of the action spectrum with the absorption spectrum shows
good coincidence. Differences are partly due to the wavelength-dependent
reflection of leaf surfaces (see reflection curve in Fig. 9.32). The spectra show
that both chlorophylls and carotenoids are involved in photosynthesis. The
action spectra do not indicate, however, whether for instance green or blue
quanta cause the same photosynthetic activity. This question is answered by
measuring a spectrum of the quantum yield. It is important that all incident
quanta are absorbed. This can be guaranteed by using a very dense algal
suspension or a thick leaf. Under these conditions we find that the
photosynthesis rate is relatively high (0.08 mol CO_2/mol quanta) even in

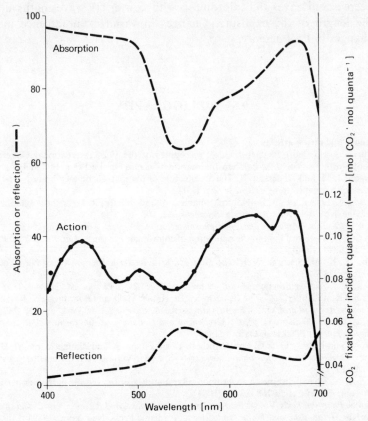

Fig. 9.32. Action, absorption and reflection spectra of a bean leaf (from Balegh and Biddulph).

green light at 520 nm, though the chlorophyll absorption is rather low in this range (Fig. 9.32). The maximal quantum yield is found at about 680 nm with 0.11 mol CO_2 per mol quanta; this value translates into 9 mol quanta per mol CO_2.

A comparison of the quantum yield of an algal suspension with that of a leaf shows a quantitative difference which can also be seen in the action spectra. Light with a high proportion of far red induces chloroplasts to produce more and thicker grana stacks, as well as an increase in PSII. The same effect is seen in shade plants which are exposed to far red-enriched light in their habitat. Since far red light is photosynthetically not very effective it has been concluded that red light causes this photomorphogenesis via the phytochrome system. Its existence has been shown in chloroplasts; it regulates some steps in thylakoid development.

Ultraviolet radiation in the UV-B, and especially in the UV-C, ranges (see Chapter 1) impairs photosynthetic activity. The variable fluorescence decreases in leaves and chloroplasts with shorter UV wavelengths allied to the length of the exposure. Changes in variable fluorescence indicate changes in PSII activity.

9.8   BIBLIOGRAPHY

Textbooks and review articles
Akoyunoglou, G. (ed.) *Photosynthesis*. Proceedings of the Fifth International Congress on Photosynthesis, Vols 1–6, Balaban International Science Service Press, Philadelphia (1981)
Anderson, J. M. and Anderson, B. The architecture of photosynthetic membranes: lateral and transverse organization. *TIBS* 288–292 (1982)
Anderson, L. E., Ashton, A. R., Habib Mohamed, A. and Scheibe, R. Light/dark modulation of enzyme activity in photosynthesis. *Bioscience* **32**, 103–107 (1982)
Barber, J. *Electron Transport and Photophosphorylation*. Elsevier, Amsterdam (1982)
Buchanan, B. B. Role of light in the regulation of chloroplast enzymes. *Ann. Rev. Plant. Physiol.* **31**, 341–374 (1980)
Clayton, R. K. and Sistrom, W. R. *The Photosynthetic Bacteria*. Plenum Press, New York and London (1978)
Cogdell, R. J. Photosynthetic reaction center. *Ann. Rev. Plant Physiol.* **34**, 21–45 (1983)
Edwards, G. and Huber, S. C. The C_4 pathway. In: Hatch, M. D. and Boardman, N. K. (eds) *The Biochemistry of Plants*, Vol. 8, *Photosynthesis*. Academic Press, New York, pp. 238–282 (1981)
Edwards, G. and Walker, D. C_3,C_4-*Mechanism and Cellular and Environmental Regulation of Photosynthesis*. Blackwell, Oxford (1983)
Gibbs, M. and Latzko, G. (eds) Photosynthesis II. In: Pirson, A. and Zimmermann, M. H. (eds) *Encyclopedia of Plant Physiology*, new series. Springer-Verlag, Berlin, Heidelberg and New York (1979)
Glazer, A. N. Phycobilisome. A macromolecular complex optimized for light energy transfer. *Biochim. Biophys. Acta* **768**, 29–51 (1984)
Govindjee *Photosynthesis*. Vol. 1: *Energy Conversion by Plants and Bacteria*. Vol. 2: *Development, Carbon Metabolism, and Plant Productivity*. Academic Press, New York (1982)
Govindjee and Jursinic, P. A. Photosynthesis and fast changes in light emission by green plants.

In: Smith, K. C. (ed.) *Photochemical and Photobiological Reviews*, Vol. 4. Plenum Press, New York and London (1979)

Hatch, M. D. and Boardman, N. K. Photosynthesis. In: Stumpf, P. K. (ed.) *The Biochemistry of Plants*, Vol. 8. Academic Press, New York (1981)

Hatch, M. D. and Osmond, C. B. Compartmentation and transport in C_4 photosynthesis. In: Heber, U. and Stocking, C. R. (eds) *Encyclopedia of Plant Physiology*, Vol. III. Springer, Berlin, pp. 144–184 (1976)

Humbeck, K., Schumann, R. and Senger, H. The influence of blue light on the formation of chlorophyll–protein complexes in *Scenedesmus*. In: H. Senger (ed.) *Blue Light Effects in Biological Systems*. Springer, Berlin and Heidelberg, pp. 359–365 (1984)

Jensen, R. G. Biochemistry of the chloroplast. In: Stumpf, P. K. and Conn, E. E. (eds) *The Biochemistry of Plants*, Vol. 1. Academic Press, New York (1980)

Junge, W., Hong, Y.-Q., Theg, S., Förster, V. and Polle, A. Localized protons in photosynthesis of green plants? In: Bolis, C. L., Helmreich, E. J. M. and Passow, H. (eds) *Information and Energy Transduction in Biological Membranes*. Alan R. Liss, New York, pp. 139–148 (1984)

Kaplan, S. and Arntzen, C. J. Photosynthetic membrane structure and function. In: Govindjee (ed.) *Photosynthesis*, Vol. 1. Academic Press, New York, pp. 65–152 (1982)

Kirk, J. T. O. and Tilney-Basset, R. A. E. *The Plastids*. Freeman, London (1976)

Malkin, R. Photosystem I. *Ann. Rev. Plant Physiol.* **33**, 455–479 (1982)

Mohr, H. and Schopfer, P. *Lehrbuch der Pflanzenphysiologie*. Springer, Berlin and Heidelberg (1978)

Ohad, I. and Klein, S. (eds) Chloroplast development: structure, function and regulation of the photosynthetic apparatus. *Israel J. Bot.* **33** (2–4), Special issue. Weizmann Scientific Press, Jerusalem, Israel (1984)

Osmond, C. B., Björkman, O. and Anderson, D. J. *Physiological Processes in Plant Ecology*. Springer, Berlin and Heidelberg (1980)

Osmond, C. B., Winter, K. and Ziegler, H. Functional significance of different pathways of CO_2 fixation in photosynthesis. In: Pirson, A. and Zimmermann, M. H. (eds) *Encyclopedia of Plant Physiology*, new series, Vol. XIIB. Springer, Berlin and Heidelberg (1982)

Raven, P. H., Evert, R. R. and Curtis, H. (eds) *Biology of Plants*. Worth, New York (1981)

Renger, G. Photosynthese. In: Hoppe, W. Lohmann, W., Markl, H. and Markl, H. (eds) *Biophysik*. Springer, Berlin and Heidelberg, pp. 532–586 (1982)

Richter, G. *Stoffwechselphysiologie der Pflanzen*. Thieme, Stuttgart (1982)

Rühle, W. and Wild, A. Die Anpassung des Photosyntheseapparates höherer Pflanzen an die Lichtbedingungen. *Naturwissenschaften* **72**, 10–16 (1985)

Salisbury, F. B. and Ross, C. W. *Plant Physiology*. Wadsworth, Belmont (1978)

Schiff, J. A. and Lyman, H. *On the Origin of Chloroplasts*. Elsevier, New York (1982)

Schilling, N. Die Photosynthese. *Prax. Naturwiss. Biol.* **29**, 97–106 (1980)

Schilling, N. Ökologische Varianten der Photosynthese. *Prax. Naturwiss. Biol.* **29**, 289–295 (1980)

Sebald, W., Schairer, H. U., Friedl, P. and Hoppe, J. On the structure and genetics of the proton-conducting F_0 of the ATP synthase. In: Bolis, C. L., Helmreich, E. J. M. and Passow, H. (eds) *Information and Energy Transduction in Biological Membranes*. Alan R. Liss, New York, pp. 175–185 (1984)

Selman, B. R. and Selman-Reimer, S. (eds) *Energy Coupling in Photosynthesis*. Elsevier, Amsterdam (1981)

Shavit, N. Energy transduction in chloroplasts: structure and function of the ATPase complex. *Ann. Rev. Plant Physiol.* **49**, 111–138 (1980)

Smith, K. C. (ed.) *The Science of Photobiology*. Plenum Press, New York (1977)

Somerville, R. and Somerville, S. C. Les photosyntheses des plantes. *La Recherche* **15**, 490–501 (1984)

Stumpf, P. K. and Conn, E. F. Lipids: structure and function. In: Stumpf, P. K. (ed.) *The Biochemistry of Plants*, Vol. 4. Academic Press, New York (1980)

Trebst, A. and Avron, M. (eds) Photosynthesis I. In: Pirson, A. and Zimmermann, M. H. (eds) *Encyclopedia of Plant Physiology*, new series. Springer, Berlin, Heidelberg and New York (1977)

Witt, H. T. Energy conversion in the functional membrane of photosynthesis. Analysis by light pulse and electric pulse methods. The central role of the electric field. *Biochim. Biophys. Acta* **505**, 355–427 (1979)

Further reading

Anderson, J. M. Consequences of spatial separation of photosystem 1 and 2 in thylakoid membranes of higher plant chloroplasts. *FEBS Lett.* **124**, 1–10 (1981)

Balegh, S. E. and Biddulph, O. The photosynthetic action spectrum of the bean plant. *Plant Physiol.* **46**, 1–5 (1970)

Barber, J. Influence of surface charges on thylakoid structure and function. *Ann. Rev. Plant Physiol.* **33**, 261–295 (1982)

Barber, J. Photosynthetic electron transport in relation to thylakoid membrane composition and organization. *Plant, Cell Environ.* **6**, 311–322 (1983)

Barnes, S. H. and Blackmore, S. Scanning electron microscopy of chloroplast ultrastructure. *Micron Microscopica Acta* **15**, 187–194 (1984)

Barsky, E. L., Gusev, M. V., Kazenova, N. V. and Samuilov, V. T. Surface charge on thylakoid membranes: regulation of photosynthetic electron transfer in cyanobacteria. *Arch. Microbiol.* **138**, 54–57 (1984)

Bassi, R., Machold, O. and Simpson, D. Chlorophyll-proteins of two photosystem I preparations from maize. *Carlsberg Res. Commun.* **50**, 145–162 (1985)

Bennett, J. Regulation of photosynthesis by reversible phosphorylation of the light-harvesting chlorophyll *a/b* protein. *Biochem. J.* **212**, 1–13 (1983)

Berry, J. A. Adaptation of photosynthetic processes to stress. *Science* **188**, 644–650 (1975)

Böhme, H. Der photosynthetische Elektronentransport und die Photophosphorylierung. *Prax. Naturwiss. Biol.* **29**, 296–302 (1980)

Bose, S. Chlorophyll fluorescence in green plants and energy transfer pathways in photosynthesis. *Photochem. Photobiol.* **36**, 725–731 (1982)

Brettel, K., Ford, R. C., Schlodder, E., Atkinson, Y. E., Witt, H. T. and Evans, M. C. W. Rapid electron transfer reaction with oxygen evolution in photosystem II preparations from spinach and *Phormidium laminosum*. *FEBS Lett.* **181**, 88–94 (1985)

Burnell, J. N. and Hatch, M. D. Regulation of C_4 photosynthesis: Catalytic phosphorylation as a prerequisite for ADP-mediated inactivation of pyruvate, P_i dikinase. *Biochem. Biophys. Res. Commun.* **118**, 65–72 (1984)

Cho, H. M., Mancino, L. J. and Blankenship, R. E. Light saturation curves and quantum yields in reaction centers from photosynthetic bacteria. *Biophys. J.* **45**, 455–461 (1984)

Cota, G. F. Photoadaption of high Arctic ice algae. *Nature* **315**, 219–222 (1985)

Dominy, P. and Williams, W. P. The relationship between changes in the redox state of plastoquinone and control of excitation energy distribution in photosynthesis. *FEBS Lett.* **179**, 321–324 (1985)

French, C. S., Brown, J. S. and Lawrence, M. C. Four universal forms of chlorophyll *a*. *Plant Physiol.* **49**, 421–429 (1972)

Glazer, A. N. Comparative biochemistry of photosynthetic light harvesting systems. *Ann. Rev. Biochem.* **52**, 125–157 (1983)

Glazer, A. N., Yen, S. W., Webb, S. P. and Clark, J. H. Disk-to-disk transfer as the rate-limiting step for energy flow in phycobilisomes. *Science* **227**, 419–423 (1985)

Govindjee, Kambara, T. and Coleman, W. The electron donor side of photosystem II: The oxygen evolving complex. *Photochem. Photobiol.* **42**, 187–210 (1985)

Ho, K. K. and Krogmann, D. W. Electron donors to P700 in cyanobacteria and algae. An instance of unusual genetic variability. *Biochim. Biophys. Acta* **766**, 310–316 (1984)

Holaday, A. S., Lee, K. W. and Chollet, R. C_3-C_4 intermediate species in the genus *Flaveria*: leaf anatomy, ultrastructure, and the effect of O_2 on the CO_2 compensation concentration. *Planta* **160**, 25–32 (1984)

Hoppe, J. and Sebald, W. The proton conducting F_0-part of bacterial ATP synthases. *Biochim. Biophys. Acta* **768**, 1–27 (1984)

Huang, C., Berns, D. S. and Guarino, D. U. Characterization of components of P-700-chlorophyll *a*–protein complex from a blue–green alga, *Phormidium luridum*. *Biochim. Biophys. Acta* **765**, 21–29 (1984)

Kyle, D. J. The 32000 Dalton QB protein of photosystem II. *Photochem. Photobiol.* **41**, 107–116 (1985)

Kyle, D. J., Kuang, T.-Y., Watson, J. L. and Arntzen, C. J. Movement of a sub-population of the light harvesting complex (LHCII) from grana to stroma lamellae as a consequence of its phosphorylation. *Biochim. Biophys. Acta* **765**, 89–96 (1984)

Ley, A. C. Effective absorption cross-sections in *Porphyridium cruentum*. Implications for energy transfer between phycobilisomes and photosystem II reaction centers. *Plant Physiol.* **74**, 451–454 (1984)

Lundell, D. J., Glazer, A. N., Melis, A. and Malkin, R. Characterization of a cyanobacterial photosystem I complex. *J. Biol. Chem.* **260**, 646–654 (1985)

Mathews-Roth, M. M. Porphyrin photosensitization and carotenoid protection in mice; *in vitro* and *in vivo* studies. *Photochem. Photobiol.* **40**, 63–67 (1984)

Melandri, B. A., Mehlhorn, R. J. and Packer, L. Light-induced proton gradients and internal volumes in chromatophores of *Rhodopseudomonas sphaeroides*. *Arch. Biochem. Biophys.* **235**, 97–105 (1984)

Ohnishi, J., Yamazaki, M. and Kanai, R. Differentiation of photorespiratory activity between mesophyll and bundle sheath cells of C_4 plants. II. Peroxisomes of *Panicum miliaceum* L. *Plant Cell Physiol.* **26**, 797–803 (1985)

Ormerod, J. G. (ed.) The phototrophic bacteria: anaerobic life in the light. In: *Studies in Microbiology*, Vol. 4. Blackwell, Oxford, London, Edinburgh, Boston and Melbourne (1983)

Renger, G. and Schulz, A. Quantitative analysis of fluorescence induction curves in isolated spinach chloroplasts. *Photobiochem. Photobiophys.* **9**, 79–87 (1985)

Robichaux, R. H. and Pearcy, R. W. Evolution of C_3 and C_4 plants along an environmental moisture gradient: patterns of photosynthetic differentiation in Hawaiian *Scaevola* and *Euphorbia* species. *Am. J. Bot.* **71**, 121–129 (1984)

Sauer, K. Photosynthetic membranes. *Acc. Chem. Res.* **11**, 257–264 (1978)

Serrano, R. Purification of the proton pumping ATPase from plant plasma membranes. *Biochem. Biophys. Res. Commun.* **121**, 735–740 (1984)

Staehelin, L. A., Golecki, J. R. and Drews, G. Supramolecular organization of chlorosomes (*Chlorobium vesicles*) and of their membrane attachment sites in *Chlorobium limicola*. *Biochim. Biophys. Acta* **589**, 30–45 (1980)

Tabata, K., Itoh, S., Yamamoto, Y., Okayama, S. and Nishimura, M. Two plastoquinone A molecules are required for photosystem II activity: analysis in hexane-extracted photosystem II particles. *Plant Cell Physiol.* **26**, 855–863 (1985)

Tanaka, A. and Tsuji, H. Appearance of chlorophyll-protein complex in greening barley seedlings. *Plant Cell Physiol.* **26**, 893–902 (1985)

Tevini, M., Iwanzik, W. and Schönecke, G. Analyse, Vorkommen und Verhalten von Carotinoiden in Kartoffeln und Kartoffelprodukten, *Forschungskreis d. Ernährungsindustrie*, pp. 36–53 (1984)

Trumpower, B. (ed.) *Function of Quinones in Energy Coupling Systems*. Academic Press, New York (1981)

Tytler, E. M., Whitelam, G. C., Hipkins, M. F. and Codd, G. A. Photoinactivation of photosystem II during photoinhibition in the cyanobacterium *Microcystis aeruginosa*. *Planta* **160**, 229–234 (1984)

Valkunas, L., Razjivin, A. and Trinkunas, G. Interaction of the minor spectral form bacteriochlorophyll with antenna and the reaction centre in the process of excitation energy transfer in photosynthesis. *Photobiochem. Photobiophys.* **9**, 139–142 (1985)

Vermaas, W. F. and Govindjee. The acceptor side of photosystem II in photosynthesis. *Photochem. Photobiol.* **34**, 775–793 (1981)

Watanabe, M., Ohnishi, J. and Kanai, R. Intracellular localization of phosphoenolpyruvate carboxykinase in bundle sheath cells of C_4 plants. *Plant Cell Physiol.* **25**, 69–76 (1984)

Westerhoff, H. V. and Dancshazy, Z. Keeping a light-driven proton pump under control. *TIBS* **9**, 112–117 (1984)

B

Processes controlled and regulated by light

10 Photomorphogenesis

10.1 INTRODUCTION

The development of an individual leads irreversibly from the embryonal stage
via the juvenile and adult stages (in which reproduction is possible) to
senescence and death. Development includes both growth and structural
differentiation. Genetically defined structural development is called morphogenesis. The morphogenesis of an organism can be influenced by external
factors, such as light, temperature, gravity, water and mineral supply. Plants
are more affected by external factors than are animals.

Light is the single most important factor, and influences biological
processes in two principally different ways. First, light supplies green plants
with the necessary energy for photosynthesis. Second, it regulates –
independent of photosynthesis – almost all stages during development. This
light-dependent control is called photomorphogenesis and comprises a large
number of photomorphogenetic reactions, the results of which are called
photomorphoses. Development in darkness is called scotomorphogenesis.

The effect of light on morphogenesis can best be demonstrated by
comparing a seedling grown in darkness with one grown in light (Fig. 10.1). In
contrast to dark-grown (etiolated) bush-bean seedlings (*Phaseolus vulgaris*),
light-grown ones have photosynthetic chlorophylls, a larger leaf area, a
shorter stem and a stronger root system. Since the etiolated and light-grown
seedlings have been reared from genetically identical seeds we must conclude
that light has regulated the genes operative during germination so that the
phenotype of the light-grown seedlings looks different from the dark-grown
ones. Light also controls subsequent development of the stem and the flowers,
as well as the senescence of the plants.

Light controls development qualitatively by inducing a certain morphosis,
such as chloroplast development in greening seedlings, which does not take
place in darkness, and/or quantatively by influencing the degree of a
genetically defined morphosis, such as growth of hypocotyls or stems. The
specific ability of a plant to respond to light during a certain developmental
stage is called competence. Many plants are very sensitive to light during their
germination. During this developmental stage the specific reactivity of various

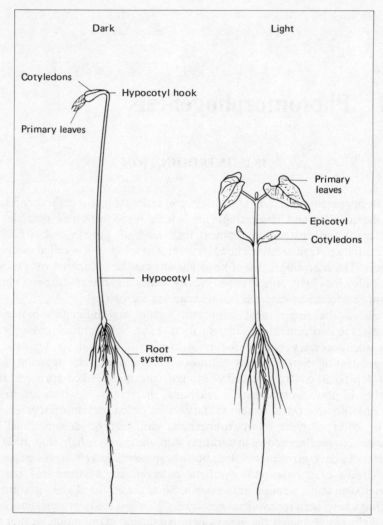

Fig. 10.1. Bush bean seedlings (*Phaseolus vulgaris*), grown in continuous light (right) and in darkness (left) (from Barber, modified).

cell types can be different, as seen for example in hypocotyls of mustard seedlings (*Sinapis alba*) in which all hypodermal cells synthesize anthocyanins while the epidermal cells do not. Obviously the endogenous biochemical condition of the cell plays an important role in the induction of a photomorphogenetic reaction.

In plants red and blue light are especially effective in inducing a photomorphogenetic reaction. In higher plants the red/far red reversible phytochrome system is the major photoreceptor pigment while in lower plants either the phytochrome system or one or more blue light/UV-A absorbing pigments (cryptochrome) play key roles in photoperception. By absorption of a quantum the photoreceptor pigment is transformed into the physiologically active form, which in turn controls cell functions and genetic expression via an unknown transduction chain, eventually resulting in a specific morphosis. The molecular basis for this effect, and the temporal and spatial coordination of the biochemical and enzymatic processes which lead to the photomorphogenesis, are largely obscure.

10.2   PHYTOCHROME SYSTEM

Photomorphogenetic reactions can be distinguished in induction reactions, which are induced by short light pulses (minutes), and continuous light responses, which need an irradiation of the order of hours (Table 10.1). Phytochrome is regarded as the photoreceptor when the reaction induced by red light can be cancelled by a subsequent far red pulse.

A classical example of photomorphogenesis induced by a short red pulse is the germination of lettuce (*Lactuca sativa*). As early as 1937, Flint and McAlister found that lettuce seeds germinate not only in white light but also in red light (< 700 nm), while far red (> 700 nm) inhibits the germination.

TABLE 10.1   PROPERTIES OF INDUCTIVE, PHYTOCHROME-REGULATED PHOTOMORPHOSES AND CONTINUOUS LIGHT MORPHOSES (AFTER MANCINELLI, EXTENDED)

	Inductive reactions	Continuous light reactions (HIR)
Fluence rates	low	high
Dose	< 200 J m^{-2}	> 2000 J m^{-2}
Reaction type	inductive	steady state
Time required	milliseconds to a few minutes	hours or days
R-FR reversibility	+	−
Action spectra (maxima)	R $= 660$ nm FR $= 730$ nm[a]	R $= 650$ nm FR 710–730 nm[b] UV-A 370 nm blue 420–480nm
Reciprocity (Bunsen–Roscoe law)	+	−
Photoreceptor pigments	phytochrome	phytochrome UV-/blue light receptor photosynthetic pigments?

[a] Mostly inhibitory.
[b] Mostly stimulatory.

TABLE 10.2   REPEATABLE R/FR REVERSIBILITY OF
GERMINATION INDUCTION OF LETTUCE SEEDS (*Lactuca
sativa* L. cv. GRAND RAPIDS). FLUENCE RATES: 1 W m^{-2}
(R), AND 5 W m^{-2} (FR) (AFTER BORTHWICK *et al.*)

Irradiation program	Germination rate (%)
R (5 min)	70
R + FR (5 min + 5 min)	6
R + FR + R	74
(R + FR)$_2$	6
(R + FR)$_2$ + R	76
(R + FR)$_3$	7
(R + FR)$_3$ + R	81
(R + FR)$_4$	7

The enhancement of germination by a 5 min red light pulse can be completely cancelled by immediately following it with a 5 min far red pulse. Another red pulse reverses the inhibition. The light quality of the last irradiation is important for induction or inhibition of germination (Table 10.2).

These results indicate that germination is controlled by a photoreceptor pigment which absorbs in either of two wavelength bands, and which can be shifted from one form into the other by different monochromatic radiation. Pigments characterized by two photoreversible states are called photochromes. The action spectra for germination induction and inhibition in lettuce seedlings (Fig. 10.2) show that the responsible pigment system has a maximum in the red spectral range at 660 nm and can be converted by red light into a far red-absorbing form with an absorption maximum at 730 nm. By a far red pulse the dark-absorbing form can be converted into the red-absorbing form.

$$P_r \underset{FR}{\overset{R}{\rightleftharpoons}} P_{fr}$$

The red absorbing form is called P_r or, because of its absorption maximum at 660 nm, P_{660}. The dark absorbing form is called P_{fr} or, due to the absorption maximum near 730 nm, P_{730}. As early as 1959, Butler demonstrated the photoreversible absorption changes at 660 nm and 730 nm in leaf homogenates of etiolated barley seedlings using a spectrophotometer specifically designed for this purpose. Phytochrome was later isolated and purified; it is a chromoprotein consisting of a chromophore and a protein. P_{fr} is assumed to be the physiologically active phytochrome form since a short pulse of weak red light induces photomorphogeneses. In fact, the degree of a photomorphogenetic effect induced by a short red light pulse can be correlated quantitatively

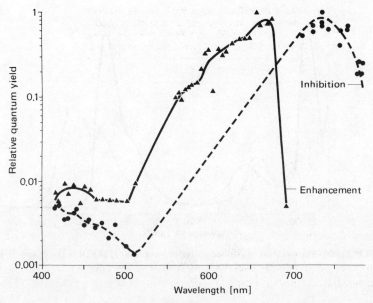

Fig. 10.2. Action spectra for photoregulation of germination in lettuce seeds (based on data by Borthwick *et al.*, from Hartmann and Haupt).

with the concentration of active phytochrome. In photomorphogeneses induced by continuous irradiation this correlation is barely found.

10.2.1   Photophysical, photobiological, and biochemical characteristics *in vitro*

Phytochrome is usually isolated from etiolated monocots such as rye, barley and corn, and from dicots such as pea, cucumber and zucchini. The absorption spectrum of purified, native barley phytochrome is shown in Fig. 10.3. The absorption maximum of the P_r form is found at 666 nm. The P_r form can be transformed into the P_{fr} form by red light which shifts the absorption maximum to 730 nm. The strong UV absorption at 280 nm is caused by the protein component. The phytochrome system can never quantitatively be converted from the P_r into the P_{fr} form since their absorption maxima overlap considerably in the red spectral region. After irradiation with red or white light we find an equilibrium between the P_r and the P_{fr} forms which depends on the spectral composition of the light source. The photostationary equilibrium (φ) at a given wavelength is defined as the ratio of the P_{fr} concentration and the total phytochrome concentration P_{tot}. These two values can be measured by

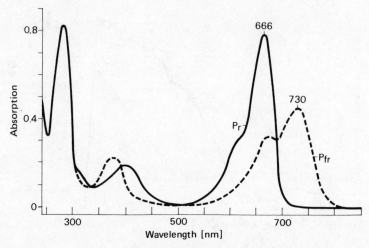

Fig. 10.3. Absorption spectrum of purified native phytochrome (124 kD) in the P_r and P_{fr} forms from etiolated *Avena* seedlings (from Quail *et al.*).

difference spectroscopy. The absorption maxima found *in vitro* correspond with those *in vivo* provided a nondegraded, native phytochrome with a molecular weight of 124,000 is used. The smaller degraded phytochrome forms (60 kD, 114 kD and 118 kD) which are due to the activity of proteases during isolation have maxima at 724 nm in the P_{fr} form.

The chemical structure of the chromophore is almost completely known but the protein structure is still obscure. The phytochrome chromophore has a molecular weight of only 612, which is about 0.5% of the molecular weight of native phytochrome (124,000). The chromophore is bound by its A ring to the amino acid cysteine in the protein. One chromophore is covalently linked to a protein monomer. It is a tetrapyrrole derivative (Fig. 10.4) and resembles the bile pigment biliverdin. In its P_r form the chromophore differs from phycocyanobilin (the chromophore of the light-harvesting pigment phycocyanin in red algae and cyanobacteria, see Chapter 9) only by an ethyl group in ring D. Because of the chemical similarity, phycocyanins and other bile pigments have been used as model systems during the identification of the phytochrome chromophore. The structural formula of the P_r chromophore has been confirmed by nuclear magnetic resonance and by a complete synthesis. The steric configuration of the substituents on ring A are not yet certain.

Recently, the chemical structure of the P_{fr} chromophore has also been revealed. The only difference to the P_r chromophore is that it is a *trans* isomer at C atom 15. During the transition from the *cis* into the *trans* form the P_{fr} chromophore tilts out of the plane by a certain angle which could expose a

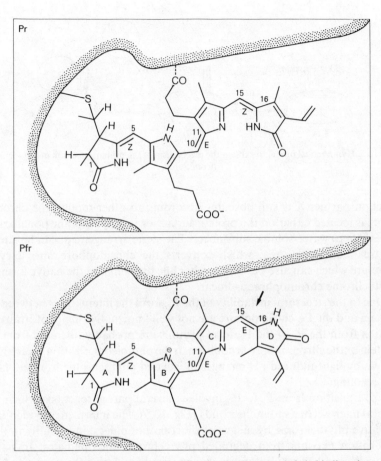

Fig. 10.4. Molecular structure of the P_r and P_{fr} chromophores and their binding sites at the phytochrome apoprotein. The steric configuration of the apoprotein has been drawn arbitrarily (from Scheer, unpublished).

binding site for a reaction partner. By binding to a partner, the P_{fr} form could be transformed into the physiologically active (effector) form. This mechanism is the basis of a model by Song (in 1980) which has been developed using data on the structural, dichroic and photometric properties of proteolytically degraded 118/114 kD phytochrome (Fig. 10.5). This model requires revision for the undegraded phytochrome molecule: In the P_r form the chromophore is buried, in the intermediate form fully exposed, and in the P_{fr} form partially exposed. In one of the intermediate states (I_{bl}?) the chromophore is most susceptible to bleaching. In the P_r form the chromophore covers a hydrophobic site of the protein surface which is exposed after the tilting of the chromophore during phototransformation. The nature of the postulated

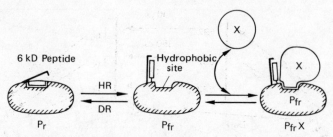

Fig. 10.5. Hypothetical model to explain the mechanism of the primary reaction of phytochrome (from Song).

reaction partner X is still obscure. According to other models the chromophore is located either on the protein surface or in a hydrophilic pouch easily accessible to water-soluble substances. The steric arrangement can be altered by treatment with urea, which converts the chromophore into a cyclic structure which can also be seen in other bile pigments. In the native form the phytochrome chromophore is linear.

Due to the structural instability of the P_{fr} form the interactions between the protein and the P_{fr} chromophore are not well known. During phototransformation from the P_r to the P_{fr} form protons are probably liberated from the protein or the chromophore in addition to the photoisomerization, as could be shown by light-induced pH changes in isolated phytochrome from etiolated pea seedlings.

The transition from P_r to P_{fr} involves several partial reactions which last several microseconds to milliseconds (Fig. 10.6). The intermediates identified in native phytochrome by means of short flashes differ considerably in their absorption maxima from degraded phytochrome intermediates. In native phytochrome the first photoreversible intermediate during the transition from P_r to P_{fr} is I_{692} (previously called lumi-R). The following reactions from I_{692} to I_{bl} and finally to P_{fr} occur in darkness with different kinetics. The first step during reversion from P_{fr} to I_{645} is light-dependent, while the following steps also occur in darkness (Fig. 10.6).

The light-independent reactions are inhibited at low temperatures or after dehydration, which indicates that the first photochemical reactions are restricted to the chromophore while the following steps involve the protein which can react only when hydrated. This result also explains the observation that seeds show only phytochrome-dependent reactions when soaked, while dry seeds do not.

Recently it has been shown that in P_{fr} not only the chromophore but also the protein is altered. In P_{fr} an additional cysteine and a histidine are easily accessible, as compared to P_r. Special amino acids seem also to be responsible for the photoreversibility. While an alternation of the cysteine

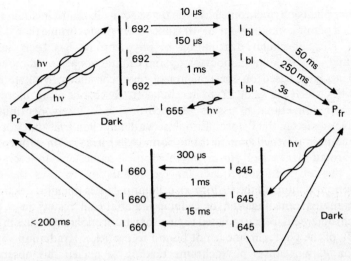

Fig. 10.6. Sequence of reactions for the transformation of native phytochrome indicating the likely intermediates, their absorption maxima and kinetics (from Furuya).

and the histidine has no effect on the photoreversibility, a chemical modification of two tyrosines in the protein completely abolishes it. This treatment destroys the structure of P_{fr}; the spectral characteristics of the P_r form, however, are retained. The conclusion is that the amino acid tyrosine is important for the stability of the chromophore in its physiologically active form.

10.2.2 Occurrence and localization

The existence of the phytochrome system has been demonstrated in angiosperms, gymnosperms, ferns, mosses, green and red algae by the red/far red antagonism of many morphogenetic and physiological reactions.

In cyanobacteria (*Tolypothrix*) a photochromic pigment system was found which shows a red/green antagonism and is called phycochrome. It is a phycobiliprotein which controls the synthesis of the blue phycocyanins and red phycoerythrins in this organism. Depending on the light quality (red or green) more blue or red pigments are synthesized. The ability to adapt to the different light conditions which prevail at different water depths is called chromatic adaptation. In addition, phycochromes control a number of other morphogenetic reactions such as cell shape, filament development and motility.

In higher plants phytochrome has been detected in all organs including roots. Etiolated plants contain about 20–100 times more phytochrome than do green plants. The subcellular distribution of phytochrome has been analyzed basically by means of immunocytology and cell fractionation. Immunocytological assay is based on the reaction of phytochrome with phytochrome antibodies which can be observed in cellular cross-sections using fluorescence microscopy. In etiolated tissues phytochrome is diffusely distributed; it exclusively exists in the P_r form. Etioplasts and mitochondria – but not nuclei – also contain phytochrome in the P_r form. After irradiation with red light (transformation into the P_{fr} form) phytochrome is sequestered to several areas within the cell, and subsequent far red irradiation restores the diffuse distribution. Binding of the P_{fr} form to cellular particles (membranes?) allows a fractionated centrifugation of cellular material and results in a higher phytochrome yield from irradiated tissues than from etiolated tissues in which 85–95% of the phytochrome is not bound to particles. Irradiation with red light for 90 min causes phytochrome binding to nuclei; this observation supports the classical concept that phytochrome controls the differential gene activity.

10.2.3   The phytochrome system *in vivo*

As *in vitro*, difference spectroscopy is also used for the quantitative analysis of phytochrome *in vivo* (Fig. 10.7). In etiolated tissues the phytochrome concentration can be determined by comparison with a reference which has been calibrated *in vitro* using purified phytochrome preparations. Often the P_{fr} fraction of the total phytochrome P_{tot} is used as a reference after irradiation with 665 nm $(P_{fr})_\infty^{665}$. The value for $(P_{fr})_\infty^{665}$ in native *Avena* phytochrome is 0.86, in other words, 86% of the total phytochrome is in the P_{fr} form and 14% in the P_r form after an irradiation with 665 nm. The concentration of P_{tot} can be calculated using these reference values:

$$[P_{tot}] = k|\Delta\Delta A|/(P_{fr})_\infty^{665}$$

where k is a proportionality factor, A the absorption difference after irradiating the sample with red and far red, respectively. $(P_{fr})_\infty^{665}$ is the phytochrome concentration (P_{fr}) in the photostationary equilibrium at 665 nm.

In green plants it is far more difficult, or even impossible, to determine phytochrome concentration and photostationary equilibrium by means of optical measurements since absorption and fluorescence of photosynthetic pigments disturb the measurement. Therefore either etiolated seedlings are used, or seedlings in which the synthesis of chlorophylls and carotenoids has

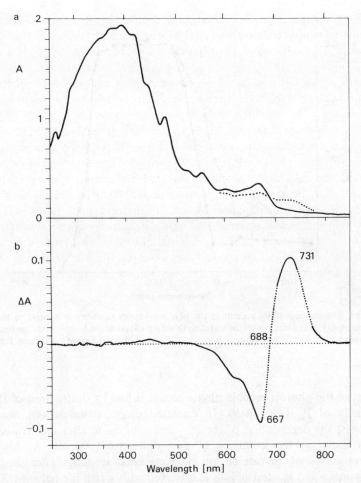

Fig. 10.7. (a) Absorption spectra of etiolated tissue from pea hypocotyl hooks before (continuous line) and after irradiation with red light (dotted line). (b) Difference spectrum of the two spectra shown in (a) (data by Inoue, from Furuya).

been inhibited by herbicides (SAN 9789 = norflurazone) which do not alter photostationary equilibrium or photomorphogenetic reactions. Figure 10.8 shows the P_{fr}/P_{tot} ratio in 54-hour-old etiolated *Sinapis* seedlings after 30 min irradiation with light of different wavelengths and fluence rates.

The total measureable amount of phytochrome in a cell depends on the *de novo* synthesis and the destruction. In darkness P_r is formed from a precursor P_r^0 and the synthesis probably resembles that of chlorophyllide since both are tetrapyrroles. Biosynthesis of the protein is thought to be under feedback control of phytochrome by regulating the level of m-RNA.

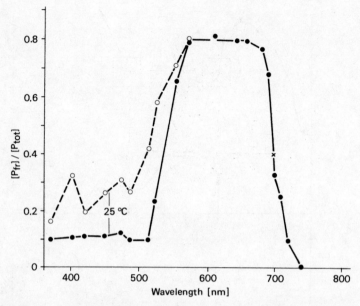

Fig. 10.8. The fraction of P_{fr} present in the photostationary equilibrium relative to the total phytochrome (P_{tot}) as a function of the wavelength in 54 h old etiolated *Sinapis* seedlings irradiated for 30 min with 1 W m^{-2} (continuous line) and 10 W m^{-2} (dashed line) at 25°C (from Jabben *et al.*).

Part of the photoreversible phytochrome is lost by destruction of P_r and especially of P_{fr} (Fig. 10.9). In etiolated dicots (*Amaranthus, Brassica, Pharbitis*) the degradation of the P_{fr} form is biphasic after red irradiation, which indicates the existence of two different P_{fr} pools. At high P_{fr} concentrations which can be found in etiolated seedlings after short red irradiation a fast P_{fr} destruction is observed with a half-life between 20 and 40 min. When this P_{fr} pool has been degraded to a large extent a slower destruction commences with a half-life of the order of hours. In green seedlings, we only find the slow destruction with a half-lifetime of about 2.5–7 h since the previous continuous irradiation has decreased the total concentration to a few per cent, which basically represents phytochrome in the stable form. Similar conditions are found in grasses.

In darkness, P_r can also be degraded provided it has previously been in the P_{fr} form. This P_r form, which is different from the initial P_r form, is thought to be synthesized in de-etiolated tissues when the P_{fr} form produced during a previous phototransformation is reverted by a far red pulse some time – for instance 1 h – later.

In addition to destruction, in some cases – for example in etiolated cucumber seedlings – a dark reversion from P_{fr} to P_r has been found to reduce

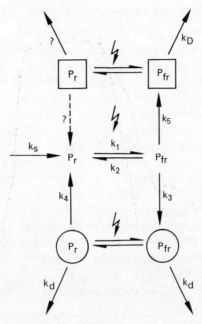

Fig. 10.9. Phytochrome conversion model including destruction (K_d and K_D), relaxation and reversion constants (k_1–k_5). k_s = time constant for the light-independent synthesis of P_r (from Schäfer).

the P_{fr} concentration. At 25°C the half-life of this thermochemical reversion is only a few minutes. Neither in etiolated nor in green dicots has a dark reversion been observed, but a dark conversion of the P_{fr} induced by red light to a modified P_{fr} form (Fig. 10.9, k_3 and k_5). The modified P_r form developed from P_{fr} relaxes in darkness to the P_r initial form with a half-life of 20–60 min (Fig. 10.9, k_4).

Based on these new results, the phytochrome conversion model (Fig. 10.9) assumes two different P_r and P_{fr} pools with different reaction constants for phototransformation (k_1 and k_2), for dark reversion (k_3–k_5) and for destruction (k_D and k_d).

The time after which the physiologically active P_{fr} concentration is reached depends on the duration of the irradiation and on whether the plant is etiolated or green because of the partially fast destruction processes. By definition we distinguish photomorphoses, which can be induced (mostly in etiolated plants) by a short red irradiation, from those which are elicited either by continuous irradiation (in etiolated plants) or, especially in nature, by daylight (in green plants). There seem to be different mechanisms by which active phytochrome controls the different reaction types, most of which are obscure.

10.2.3.1 INDUCTIVE PHOTOMORPHOGENESES

Some of the best-known examples of inductive photomorphogeneses are the germination of lettuce seeds (*Lactuca sativa*) as well as the enhancement of coleoptile and inhibition of mesocotyl growth (see Section 10.4.2) in etiolated barley seedlings (*Avena sativa*). Short red irradiation can also induce biochemical photomorphogeneses, such as anthocyanin and flavonoid synthesis, though the maximal production of these pigments occurs only under long-term irradiation (HIR, see below).

The key enzyme for the synthesis of secondary plant substances is phenylalanine ammonium lyase (PAL), the synthesis of which is probably induced by P_{fr} by means of differential gene activation. PAL deaminates phenylalanine to *trans* cinnamic acid, which is the precursor for the biosynthesis of the B ring of flavonoids and anthocyanins. The anthocyanins in the vacuoles of the subepidermal cells of mustard seedlings are responsible for the red color and are the macroscopically visible result of a biochemical photomorphogenesis (Fig. 10.10).

The Bunsen–Roscoe law (law of reciprocity, see Section 12.2.1.2) holds for

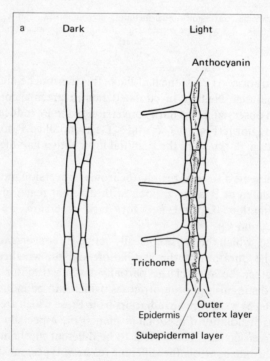

Fig. 10.10a.

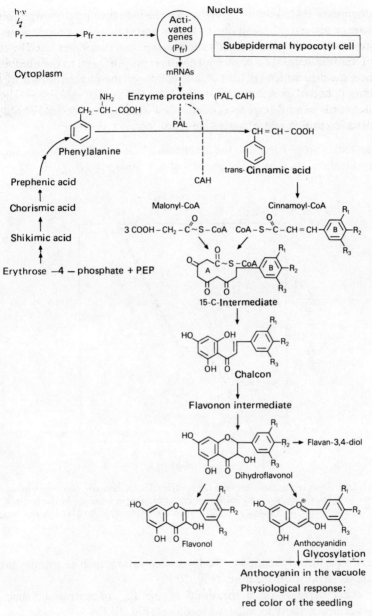

Fig. 10.10. Effect of light (phytochrome system) on the cell morphology and anthocyanin synthesis in mustard seedlings. (a) Photomorphogenetic changes in the outer cell layers of the hypocotyl (schematic longitudinal section). (b) Phytochrome induction of anthocyanin synthesis (from Mohr and Sitte, modified).

the induction of PAL activity as well as for the inductive photomorphoses of germination and growth. Simplified, this law states that the effect depends on the doses, and that fluence rate and exposure time can be compensated by each other. Thus, the degree of a photomorphosis is proportional to the number of absorbed quanta, which in turn is a function of the concentration of the absorbing pigment, its spectral properties and the fluence rate.

Inductive photomorphogeneses of etiolated seedlings can depend on the P_{fr} concentration in different quantitative relations:

1. The effect is proportional to the logarithm of the radiation doses or the P_{fr} concentration, such as in the growth of oat coleoptiles (Fig. 10.11).

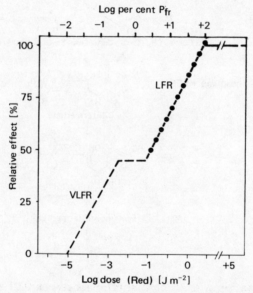

Fig. 10.11. Log–linear relationship between the relative effect of inductive photomorphoses and the fluence (red, lower abscissa) and the P_{fr} concentration (upper abscissa) in the range of the VLFR (very low fluence response) and LFR (low fluence response) (from Mandoli and Briggs).

2. The effect is linearly related to the P_{fr} concentration, such as anthocynanin synthesis in mustard seedlings (Fig. 10.12).
3. The effect depends on a threshold of the P_{fr} concentration, such as lipoxygenase activity in mustard seedlings (Fig. 10.13).

The dose–response curve (Fig. 10.11) generalized for a number of inductive photomorphogeneses is biphasic. The first phase is called VLFR (very low fluence response), the second LFR (low fluence response). There is a horizontal, dose-independent section between the two. The VLFR follows a

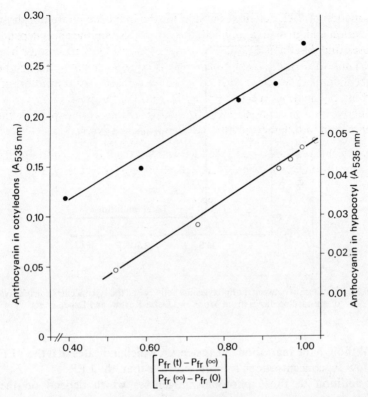

Fig. 10.12. Linear relationship between the P_{fr} content and the anthocyanin content (absorption at 535 nm) in cotyledons (closed circles) and in the hypocotyl of mustard seedlings (open circles) (from Steinitz *et al.*).

log-linear relationship with P_{fr} concentrations between 0.01 and 0.4%. The LFR starts at about 2% P_{fr} and is saturated at high P_{fr} concentrations. The green 'safe light' which is often used in experiments can already induce the first part of the biphasic curve, unnoticed by the experimenter. The biphasic curve can be interpreted as follows: an extremely low P_{fr} concentration induces transport of an effector through the membrane. This effector activates an enzyme which degrades P_{fr}. In the horizontal part of the curve P_{fr} synthesis and P_{fr} destruction balance each other at certain irradiation doses (about $0.01–0.1 \, J \, m^{-2}$). Increasing doses saturate the enzyme activity and the P_{fr} concentration increases. As a consequence we find an exponential relationship between the dose and effect of the second phase.

In contrast, the amount of synthesized anthocyanin in the hypocotyl and cotyledons of mustard seedlings depends linearly on the P_{fr} concentration; this relationship is valid for the LFR above 40% P_{fr} (Fig. 10.12) as well as for the

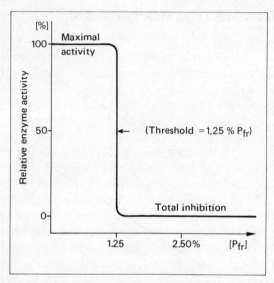

Fig. 10.13. Relationship between the lipoxygenase activity and the P_{fr} concentration in cotyledons of mustard seedlings (from Mohr and Oelze-Karow, and Lange *et al.*).

VLFR though the regression line is steeper, which indicates that the VLFR at very low P_{fr} concentrations is more sensitive than the LFR.

In addition to those photomorphogeneses which depend on the P_{fr} concentration by some linear or logarithmic function, some photomorphogeneses are all-or-none events (Fig. 10.13). Such a relationship is found for the enzyme lipoxygenase, which catalyzes the oxidation of polyene fatty acids, such as arachidonic and linoleic acid which have a *cis,cis*-1,4-polyene structure, to conjugated *cis,trans*-dienehydroperoxides in fat-containing seeds and cotyledons. The lipoxygenase is inhibited at a P_{fr} concentration of more than 1.25% and active below. In darkness this threshold is not reached, and the enzyme is active.

Many investigations, especially of phytochrome-controlled anthocyanin synthesis, indicate that etiolated seedlings are able to measure the absolute amounts of P_{fr} and use it as a control value. In some cases, however, there is no quantitative correlation between the effect and the control value of P_{fr}. Some of these exceptions are so-called 'paradoxes' such as the *Zea* paradox where a certain effect is induced by certain red doses which are well below those necessary to induce a spectrophotometrically measurable phototransformation of P_r to P_{fr}. In addition, this effect is reversible by far red doses which induce a higher P_{fr} concentration in the photostationary equilibrium than those red doses necessary to saturate the effect. As for the VLFR we find an extremely high sensitivity in the low P_{fr} range.

10.2.3.2   CONTINUOUS LIGHT MORPHOGENESES

The amount of anthocyanin synthesized after light induction is far lower than that produced after several hours of irradiation with red or far red light at high fluence rates. In mustard seedlings far red is more effective than red light, since it induces the same anthocyanin synthesis at lower fluence rates than those necessary with red light. For both light qualities the amount of anthocyanins produced depends on the irradiation time (Fig. 10.14): a far red irradiation at 0.35 W m^{-2} for 12 h induces a considerable anthocyanin synthesis, while a far red irradiation at 3.5 W m^{-2} for 1.2 h is completely ineffective though both treatments amount to the same dose of 15,120 J m^{-2}. Photomorphogenic reactions which require a multiple of the exposure time needed for a response with inductive character (at the same fluence rate) are called high irradiance responses (HIR). Typical examples of HIR are the responses of etiolated seedlings which require a continuous irradiation. Green plants reared in artificial or daylight also show HIR; they differ, however, in their dependence on the concentration of P_{fr}. Though the degree of a HIR photomorphosis is a function of both fluence rate and exposure time, reciprocity does not hold for HIR: low fluence rates applied over a long time are more effective than high fluence rates irradiated for a short time. The reasons for this phenomenon could be that the photoreceptor pigment concentration changes with the

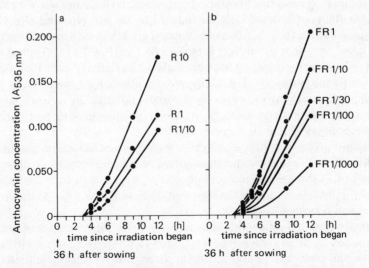

Fig. 10.14. Anthocyanin accumulation in mustard seedlings under continuous (a) red or (b) far red light with different fluence rates in dependence of the irradiation time. The fluence rate of R1 was 675 mW m^{-2}, that of FR1 was 3.5 W m^{-2}; the other fluence rates were multiples or fractions thereof (from Lange *et al.*).

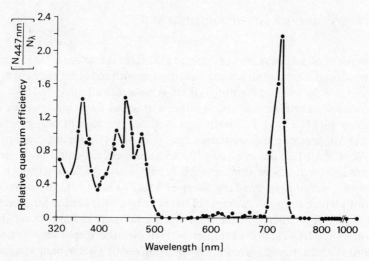

Fig. 10.15. Action spectrum of the inhibition of hypocotyl growth of lettuce seedlings (*Lactuca sativa*, cv. Grand Rapids) by continuous light in the time interval between 54 h and 72 h after sowing (from Hartmann).

fluence rate, or that the tissue becomes more sensitive toward the effector during irradiation.

In contrast to inductive photomorphogeneses, HIR do not show a red/far red reversibility. 'Classical' HIR are induced by far red, blue and ultraviolet radiations, such as the inhibition of hypocotyl growth in lettuce seedlings (Fig. 10.15). Red as well as green light is barely effective (Fig. 10.16). Typical HIR occur only when the seedling has accumulated a relatively high P_{fr} level. In addition to the 'classical' HIR, different continuous light phenomena have been found, which are elicited by far red and blue light, such as the anthocyanin synthesis in cabbage or the hypocotyl growth inhibition in *Sinapis* seedlings (Table 10.3, group 1).

A second group of continuous light morphogeneses has action maxima in the ultraviolet, blue and red spectral regions, such as anthocyanin synthesis in apple skins. A third group includes photomorphogeneses which are induced only by continuous ultraviolet or blue radiation, such as anthocyanin synthesis in *Sorghum* seedlings.

In contrast to inductive photomorphoses, the action spectrum of growth inhibition in etiolated lettuce hypocotyls shows a pronounced sensitivity to far red, UV and blue radiations (Fig. 10.15). Growth is specifically inhibited by prolonged far red radiation, while red as well as green light is almost ineffective (Fig. 10.16). Since the hypocotyl grows linearly with time, the inhibition by far red is equal at all times. However, the inhibition is not proportional to fluence

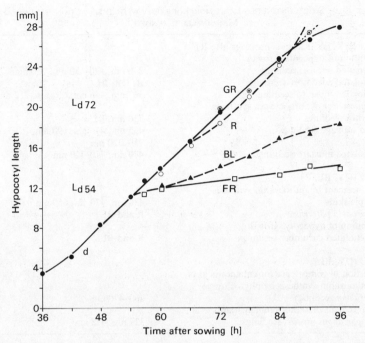

Fig. 10.16. Growth curves of lettuce hypocotyls at $25 \pm 0.5°C$ (*Lactuca sativa* L. cv. Grand Rapids). Conditions: d = Continuous darkness. BL, GR, R and FR: Irradiation starting 54 h after sowing using blue (450 nm, HW = 43 nm), green (530 nm, HW = 30 nm), red (658 nm, HW 15 nm) or far red (720 nm, HW = 120 nm). The photon fluence rate was 6.6 μmol m^{-2} s^{-1} in blue, green and red and 46.3 μmol m^{-2} s^{-1} in far red light, respectively. The additional lines indicate the best irradiation time interval (54 to 72 h) as well as the mean initial and final length of the hypocotyls in darkness (L_{d54}, L_{d72}) (from Hartmann).

rate but increases logarithmically; thus reciprocity does not hold for this case either.

These results indicate that HIR are photocatalytic reactions. Unlike many inductive photomorphogeneses, HIR do not depend only on the P_{fr} concentration, but on an additional component X which reacts with phytochrome and catalytically affects the reaction (see Section 10.2.4). It could be shown by simultaneous irradiation with two different wavelengths that the catalyst is phytochrome itself, and that both P_r and P_{fr} need to be excited to induce the effect. Simultaneous irradiation of two relatively inefficient wavelengths (658 and 768 nm) induce an optimal effect when a low but effective phytochrome equilibrium is induced (Fig. 10.17). The receptor sites are saturated with the partner X up to a value of 3% active phytochrome, which induces optimal photomorphogenetic reactions. Even a further increase in the P_{fr} concentration by higher red doses would not change the P_{fr}–X status and the reaction would continue normally. However, this is not the case, as

TABLE 10.3   SOME CONTINUOUS LIGHT MORPHOGENESES WITH THEIR ACTION MAXIMA (AFTER
MANCINELLI, EXTENDED)

Group 1 (UV, BL, FR, in some cases also R)
  Inhibition of hypocotyl growth

etiolated lettuce seedlings	370 nm, 420–480 nm, 720 nm
etiolated cucumber seedlings	R, DR, BL
etiolated mustard seedlings	367 nm, 446 nm, 653 nm, 712 nm

  Enhancement of anthocyanin synthesis

turnip seedlings	730 nm, BL
red cabbage seedlings	435 nm, 470 nm, 690 nm, 710–720 nm
etiolated mustard seedlings	470 nm, 710–720 nm

Group 2 (UV, BL, R)
  Enhancement of anthocyanin synthesis

apple skins	430 nm, 470 nm, 650 nm
Spirodela polyrrhiza	R and BL

  Inhibition of hypocotyl growth

deetiolated cucumber seedlings	R and BL

Group 3 (UV, BL)
  Induction of competence for enhancement of
  anthocyanin synthesis by phytochrome

Sorghum seedlings	460–470 nm

  Enhancement of anthocyanin synthesis

Haplopappus gracilis seedlings	327 nm, 438 nm

has been shown experimentally by simultaneous irradiation with two wavelengths. Additional red light reduces the inhibitory effect on hypocotyl growth. Obviously, additional reactions occur at higher P_{fr} concentrations, such as an increased P_{fr} destruction or a change in the reaction partner X which prevents a lasting and effective combination of P_{fr} and X.

Anthocyanin accumulation in continuous far red light can also be explained by an effective P_{fr} concentration. Reciprocity does not hold for this case either, since the synthesized amount of anthocyanin is higher after a 30 h far red irradiation with a fluence rate of 0.25 W m^{-2} than after 3 h ($+27$ h darkness) at a fluence rate of 2.5 W m^{-2}. The 30 h far red irradiation with a low fluence rate induces a constant P_{fr} concentration which amounts to 2–3% of P_{tot}. Due to the dark reversion and dark destruction of P_{fr} in the dark period following the 3 h irradiation there is no effective P_{fr} concentration left.

The nature of the photoreceptor pigment for continuous light reactions is still obscure. In addition to the phytochrome system, a yet unknown blue light receptor may be involved, since most continuous light reactions have a second action maximum in the blue spectral range. The action spectrum for growth inhibition in the lettuce hypocotyl has three peaks in the blue range and

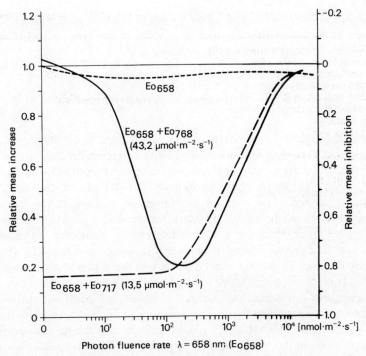

Fig. 10.17. Growth inhibition of the lettuce hypocotyl by dichromate irradiation. Continuous line: addition of almost completely ineffective red light with increasing photon fluence indicated on the abscissa ($=E_{o\,658}$) to a constant photon fluence of ineffective far red irradiation ($=E_{o\,768}$) varies the relative mean inhibition between 0 and 0.8. Dashed line: inhibition caused by far red disappears with increasing addition of red light ($=E_{o\,658}$). Short dashes: red control ($=E_{o\,658}$) (from Hartmann, modified).

resembles the absorption spectrum of carotenoids. Action spectra of other high irradiance responses have only one peak in the blue range and resemble flavin absorption. Therefore both flavins and carotenoids are discussed as possible photoreceptor candidates; flavins are currently favored for several reasons (see Section 10.3).

The following results suggest the existence of an independent blue light receptor pigment. In *Sorghum* a long-lasting blue light irradiation is the prerequisite for the subsequent control of anthocyanin synthesis by phytochrome. Mutants of *Arabidopsis thaliana*, which have much less phytochrome than the wild type, have lost the inhibition of hypocotyl growth by red light but not by blue light. A flavin-induced increase of the P_r to P_{fr} photoconversion has been observed *in vitro*, which also indicates the activity of a separate blue light photoreceptor. Currently the identity of the blue light receptor responsible for these reaction is as obscure as that involved in exclusively blue light controlled morphoses (see Table 10.5). In addition, we should keep in mind that both phytochrome forms absorb in the blue spectral range, and it cannot be excluded that phytochrome itself is involved in the blue light absorption in many photomorphogeneses.

10.2.4   Mechanisms of phytochrome action

We can distinguish two different action mechanisms controlling photomorphogeneses: a reversible modulation and an irreversible differentiation (Table 10.4). The leaf movements of *Albizzia*, for example, are controlled by reversible modulation. These photomorphogeneses are based on the control of physiological reactions, such as ion fluxes or changes in the membrane potential or the permeability. However, there are also growth processes and enzyme activities which are modulated reversibly. Examples of this mechanism are the hypocotyl growth of several seedlings and the ATP synthase activity in beans. Due to the short time span between stimulus and response a genetic control can be excluded.

Irreversible differentiations, such as flower induction or anthocyanin synthesis, are controlled by phytochrome, probably via a regulation of the gene expression. This relatively old hypothesis was recently supported by measurements of the translation activity of mRNA *in vitro*, which showed a phytochrome-induced decrease of mRNA activity for the protochlorophyll reductase and an increase in the *de novo* synthesis of the small subunit of ribulose-bisphosphate carboxylase. According to recent results, phytochrome even controls the transcription or translation of its own apoprotein gene.

Both modulations and irreversible photomorphoses include a signal transduction between the P_{fr} formation and the morphogenetic reaction which

TABLE 10.4   SOME EXAMPLES FOR THE EXPRESSION OF PHYTOCHROME-CONTROLLED PHOTOMORPHOGENESES

Reversible modulations
   chloroplast movements in *Mougeotia*
   leaf movements in *Albizzia*
   surface and membrane potentials of *Avena* coleoptiles
   acetate uptake in *Phaseolus*
   inhibition of shoot growth in *Phaseolus* and *Pisum*
   inhibition of hypocotyl growth in *Sinapis*
   enzyme activity (ATPase) in *Phaseolus*

Irreversible differentiation
   induction of seed germination in *Lactuca*
   induction of hypocotyl hook opening in *Sinapis*
   induction of cotyledon (*Sinapis*) and leaf growth in *Phaseolus*
   flower induction in *Hyoscyamus*
   enzyme induction (PAL) and regression (lipoxygenase) in *Sinapis*

is a translation of the light-induced signal into a sequence of molecular events. The transduction chain starts with the primary reaction of P_{fr} with one or more still unknown, possibly membrane-bound, receptors. The subsequent reactions leading to the expression of the photomorphogenetic reaction are also obscure. Phytohormones may play a role in the signal chain. It could be shown, for instance, that gibberellic acid stimulates leaf unfolding in monocot seedlings, as does red light. It is not clear, however, how phytohormones operate within the reaction chain.

The transduction is a fast event (within minutes) and the reaction can either also take place within minutes or it is delayed for several hours or even days. In the second case the fast P_{fr} responses can be deduced from the time within which a red/far red reversion is possible. In most of the typical photomorphoses the manifestation is delayed for a relatively long period of time, which indicates that P_{fr} has to be effective either continously or at least during several critical steps in the signal chain.

Figure 10.18 shows a few hypothetical possibilities of how active phytochrome can start the signal transduction. The first model (Schäfer, 1975; Mohr, 1978) assumes the interaction of P_{fr} with two hypothetical receptors X and X' (Fig. 10.18a). Irradiation induces the formation of $P_{fr}X$ which converts in darkness into the stable form $P_{fr}X'$. $P_{fr}X$ and $P_{fr}X'$ are photoreversible and $P_{fr}X'$ can relax to either P_rX or P_r. This model suggests that in continuous light reactions soluble phytochrome accumulates as $P_{fr}X$, which is thought to be the physiological effector. In darkness the $P_{fr}X$ produced rapidly converts into the stable $P_{fr}X'$, which is then the only physiological effector. The primary reaction of the $P_{fr}X$ forms leads to the formation of the product Y, which induces the photomorphogeneses. Probably not all photomorphogeneses are controlled by the same signal transduction chain. It is possible that the different reaction chains are already selected on the level of the primary reactions and their products. During the primary reaction P_{fr} can cooperate with the reaction partner X. This seems to be the case with the threshold reaction of lipoxygenase (see Section 10.2.3.1). The switching from active to inactive state within a very narrow threshold range can be explained only by a high cooperation between P_{fr} and X. The reaction partner X is believed to be a membrane component, and to induce a different conformation of the membrane in connection with P_{fr} than with P_r. The fact that membrane-bound P_r and P_{fr} sediment differently supports this hypothesis.

In an alternative model (Smith, 1970) phytochrome is assumed to induce the transport of metabolites or phytohormones X^0 through a membrane (Fig. 10.18b). The transport is mediated by the photoconversion of the phytochrome forms. After transport X^0 reacts with another metabolite which induces the photomorphogenetic processes.

A similar model was developed by Johnson (1979) (Fig. 10.18c). The basic difference to Smith's model is the reaction of X^0 with P_{fr}, rather than with a

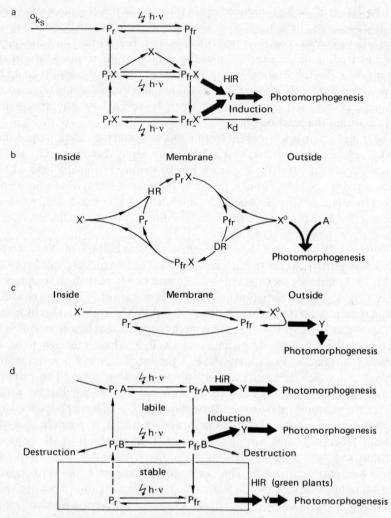

Fig. 10.18. Various hypothetical models to explain the mechanism of the phytochrome system. (a) Model from Schäfer and Mohr, (b) model from Smith, (c) model from Johnson, (d) model from Jabben and Holmes.

metabolite A, which induces the reaction. Under inductive conditions the reaction therefore depends on the P_{fr} concentration (with the exception of the paradoxes) and not on X^0, the accumulation of which is a function of the cyclic photoconversion rate and the number of phytochrome molecules. Since P_{fr} is steadily destroyed there is always less P_{fr} than X^0. Under HIR conditions, when P_r is continuously photoconverted to P_{fr}, there is a constant P_{fr} concentration and the level of X^0 determines the reaction.

Recently Jabben and Holmes (1983) have developed a model for etiolated and green plants which assumes the existence of a labile and a stable phytochrome pool (Fig. 10.18d). In etiolated plants we find predominantly the labile pool which contains the effector forms $P_{fr}A$ and $P_{fr}B$, which correspond with $P_{fr}X$ and $P_{fr}X'$ in Schäfer's model. HIR are suggested to be induced by $P_{fr}A$ and inductive photomorphogeneses by $P_{fr}B$. In green plants the stable phytochrome pool predominates and P_{fr} is the only effector. P_{fr} reacts with specific binding partners which may be different from the binding partners of other phytochrome forms. This model explains the long-known phenomenon that continuous light responses (HIR) in many green plants are maximally induced by red light, while in etiolated plants far red induces maximal activity.

10.3   BLUE LIGHT-DEPENDENT PHOTOMORPHOGENETIC REACTIONS

Blue light-dependent morphogenetic reactions are common among algae, fungi and ferns; they are found in higher plants and animals as well. In the brown alga, *Scytosiphon*, blue light causes a two-dimensional growth, while red light induces a filamentous growth (Fig. 10.19a, b). Similar conditions are found in the fern *Dryopteris filix-mas*, which forms a normal prothallium in blue light and only a filamentous chloronema in red light. In fungi photomorphogenetic responses are predominantly induced by blue and ultraviolet radiation such as perithecia development in *Nectria* or conidia induction in *Trichoderma*. The action spectra usually show three peaks in the blue spectral range and a secondary peak at about 365 nm. It is interesting to note, however, that in some fungi exclusively UV radiation induces sporulation. Due to the different action maxima in the long and short wavelength UV and blue spectral range the activity of several photoreceptor pigments is feasible. The identity of a photoreceptor pigment responsible for the action in the blue and UV ranges is still unknown, and it is therefore called cryptochrome. The same receptor seems to be responsible for phototropism in shoots and coleoptiles which is induced by blue but not by red light (see Section 12.2.1). In the blue range the action spectrum for hypocotyl growth inhibition in lettuce seedlings (Fig. 10.15) resembles that of the first positive phototrophic bending in *Avena* coleoptiles. For hypocotyl growth inhibition by red light, however, phytochrome is responsible (HIR).

Blue and red light effects can be separated temporally and spatially, which supports the hypothesis of two separate receptor systems. Hypocotyl growth in cucumber seedlings is inhibited by blue light within 5 min, but by red light only after 45 min. In addition, we find a spatial separation of the receptor

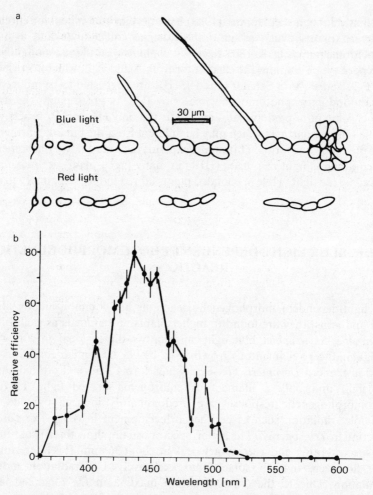

Fig. 10.19. (a) Schematic diagram of the morphological development of *Scytosiphon lomentaria* in blue and red light. (b) Action spectrum for the two-dimensional growth of *Scytosiphon lomentaria* (from Dring and Lüning).

systems in greening cucumber hypocotyls: blue light is perceived directly by the hypocotyls while red light is perceived by the cotyledons. However, there is a number of other examples of the cooperation of the blue light/UV receptor with the phytochrome system. It could be shown that phytochrome stimulates anthocyanin synthesis in mesocotyls of *Sorghum* seedlings only after a blue light pretreatment. The same situation is found in the UV-induced flavonoid synthesis in parsley cell suspension cultures.

Numerous blue light effects have been described for plastid development, such as chloroplast development in roots or cell cultures. The effect of blue light on protein synthesis during greening of *Euglena* could be demonstrated on the levels of transcription and translation. The blue light/UV receptor system has neither been isolated nor chemically characterized.

The action spectra of many blue light-mediated reactions can be explained by assuming either flavins or carotenoids as photoreceptor pigments. The arguments in favor of the two alternative receptors are summarized in Table 10.5. A number of action spectra have peaks at 450 and 370 nm, which resembles the flavin absorption spectrum; other action spectra lack the 370 nm maximum and resemble a carotenoid spectrum. Therefore the blue light-inhibited circadian conidia development in *Neurospora*, which lacks the UV inhibition, is believed to be due to carotenoids. However, *Neurospora* mutants, which only have traces of carotenoids, still respond to blue light. This result favors flavins over carotenoids unless extremely small carotenoid concentrations could be morphogenetically active.

Strong arguments in favor of flavins as photoreceptor pigments are found in their photochemistry. Almost all blue light-dependent reactions require the presence of oxygen, which indicates a photooxidation during the primary steps of the blue light response. In fact, the blue light-dependent carotenoid synthesis in *Fusarium* could be stimulated by hydrogen peroxide. Furthermore, blue light effects can be suppressed by reductants such as dithionite. Addition of photodynamically active substances (photosensitizers) such as methylene blue also showed that some blue light reactions are redox reactions which can be performed by flavins but not by carotenoids. Adding flavin solutions to cell-free systems caused membrane-bound cytochromes to show redox reactions in blue light. These light-induced absorbance changes (LIAC)

TABLE 10.5   ARGUMENTS FOR FLAVINS AND CAROTENOIDS AS POSSIBLE BLUE LIGHT RECEPTORS (AFTER SCHMIDT)

Pro flavins
  UV maximum between 350 nm and 400 nm in the action spectra
  O_2 dependence of the primary reactions
  flavin reactions are often redox reactions
  oxidants can substitute light, reductants suppress the blue light reaction
  flavin inhibitors (e.g. KI) inhibit the blue light reaction
  low-temperature spectra of flavins resemble the blue light action spectra
  blue light action in carotenoid free *Neurospora* mutant
  lifetime of carotenoids in the first excited singlet state is very short (10^{-13} s)

Pro carotenoids
  three peaked action spectra
  no or small UV maximum in some action spectra
  energy transfer from a UV absorbing pigment to carotenoids feasible
  no blue light action in carotenoid-free diatom mutant

of cytochromes have recently been demonstrated in isolated plasma membrane particles.

Figure 10.20 shows the possible excitations of a flavin in a Jablonski diagram. In most cases we find π, π^* transitions. Singlet states of carotenoids have such short lifetimes (10^{-14} s) that they cannot be responsible for the induction of photochemical reactions. For energetic reasons the direct triplet excitation of carotenoids is rather unlikely, but long-lived triplet states could be induced in carotenoids by energy transfer from another triplet donor.

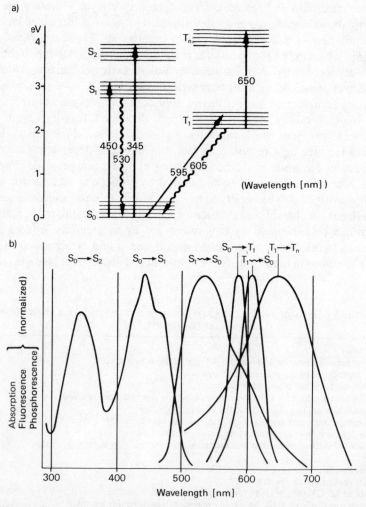

Fig. 10.20. (a) Jablonski diagram for the flavin chromophore. (b) Various electronic transitions for the flavin nucleus (normalized spectra from Schmidt).

10.4 PHOTOMORPHOGENESIS IN PLANTS

In plants growth is characterized by an irreversible increase in cell number and cell size. The increase in cell number is effected by genetically controlled mitotic divisions of meristematic cells. The differentiation of cells and tissues is based on the differential activation and inactivation of genes. The control and regulation of gene activity depends both on autoregulatory control systems and on control genes which are regulated by external factors such as light. Unicellular fern spores are a simple and instructive example: in lateral light they grow green thallus cells towards the light and colorless rhizoids away from the light source. During the first cell division the chromosomes are equally distributed to the daughter cells but not the cytoplasm, which is polarly arranged. Thus, light induces the polarity of the fern spore and dictates the direction of the unequal cell division.

The stability of the polar cytoplasm distribution depends on the photo-control of enzymes. Light can either activate or inactivate enzymes directly or influence the *de novo* synthesis. The nitrate reductase in bean leaves, for example, is activated, the peroxidase in cucumber hypocotyls is inhibited and the *de novo* synthesis of aminoacyl-tRNA synthetase in pea leaves is enhanced by light. The competence of enzymes to respond to light depends on the age of the plant. In *Sinapis* seedlings the lipoxygenase is under phytochrome control only after 33 h, and 15 h later the phytochrome control disappears again. The competence for the photocontrol seems to be due to endogenous control mechanisms on the level of translation and/or transcription. In mustard cotyledons the phytochrome system stimulates the synthesis of the cytoplasmic and plastidic rRNA, possibly via a DNA-dependent RNA polymerase. In greening barley seedlings light-induced changes have been found in the polysomal mRNA.

10.4.1   Germination

Germination can be induced by light in soaked seeds of many plant species. We distinguish plants, the seeds of which germinate in light (for example *Lactuca sativa* (see Fig. 10.2 and Table 10.2), *Daucus carota*, *Sinapis alba*, *Rumex crispus*) from those which germinate in the dark (for example *Amaranthus caudatus*, *Lycopersicum esculentum*, *Cucurbita pepo*) and in which light inhibits germination. *Phacelia tanacetifolia* also belongs to the latter group; in this plant germination is even inhibited in darkness after a far red pretreatment. In many plant species the activation of seeds depends on the length of the daily light period. Plants which germinate in long days are, for example, *Betula pubescens* and those which germinate in short days *Veronica persica*.

The first visible morphological process during germination of lettuce seeds is the growth of the radicula and the hypocotyl. Initially the growth of the radicula cells depends exclusively on cell enlongation due to an increase of the osmotic potential and a decrease of the wall pressure by loosening the cell wall. Both processes can be elicited by red light in embryos from which the testa has been removed. The mechanical properties of the endosperm play an important role during germination since in some cases it prevents the passage of the radicula. Due to the effect of P_{fr} on the synthesis or activation of enzymes, such as mannanase which degrades the reserve substance mannan to sugars, the endosperm softens. By removing the endosperm in *Phacelia*, which usually germinates in the dark, germination can be induced by light.

The ability to germinate depends on the P_{fr} concentration: the higher the P_{fr} concentration the higher the percentage of germinating seeds, as could be shown for *Sinapis arvensis* seeds. The P_{fr} concentration is controlled by adjusting the red/far red ratio of the light source. The P_{fr} concentration in the photoequilibrium calculated from the spectral properties of the light source is in good agreement with the experimentally measured values in the same irradiation conditions (Fig. 10.21).

Plant species differ in the time period during which active P_{fr} has to be present to induce maximal germination. While *Sinapis* needs only 15 min red light for maximal germination, *Plantago major* requires 48 h. In *Sinapis arvensis* the period is also shorter during which the red light effect can be reversed. *Sinapis* looses its photoreversibility within 3 h, while *Plantago* retains its reversibility for about 24 h (Fig. 10.22). However, the time period during which the effect can be reversed strongly depends on the ambient temperature: reversibility is lost faster at 25°C than at 15°C. It can be concluded that the temperature affects the formation of the primary product either by changing the phytochrome receptor site or by altering the speed of biochemical reactions which are catalyzed by P_{fr}.

We have mentioned that active phytochrome induces and stimulates germination in light-dependent seeds. We could conclude that no germination occurs in the dark since we usually find only the more stable P_r form in seeds. However, Fig. 10.22 shows that *Sinapis* seeds, in contrast to *Plantago* seeds, have a high germination rate of 20–40% in the dark. Probably *Sinapis* seeds contain some P_{fr} in the dark as do many other dark-germinating seeds. In many cases the phytochrome form present in the seeds depends on the preirradiation of the mother plant. In *Arabidopsis* the light quality with which the mother plants has been irradiated controls the germination ability of the seeds. When the light contained a high proportion of far red the plants formed seeds with phytochrome predominantly in the P_r form. When the light contained barely any far red most of the phytochrome was in the P_{fr} form and

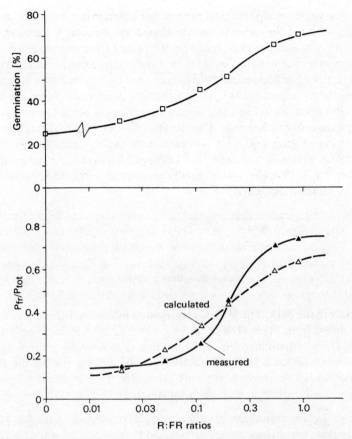

Fig. 10.21. Relationship between the germination rate of *Sinapis arvensis* seeds and the $P_{fr}:P_{tot}$ ratio in dependence of the red:far red ratio of the light source (from Waddoups).

the seeds germinated in the dark. In lettuce seeds it could be shown that when the phytochrome was tranformed into the P_{fr} form, and the seeds were dehydrated immediately afterwards, they germinated in the dark even after prolonged storage. In dry seeds a phytochrome conversion is not possible, since the conversion requires the phytochrome protein to be hydrated. The intermediate reactions which involve only the chromophore (see Fig. 10.6) can, however, occur in dry seeds and the intermediates can be converted back to the initial form in darkness.

Generally, the inhibition of germination by long-time far red is found in seeds germinating in the dark, such as *Phacelia tanacetifolia* and *Amaranthus caudatus*, but it is also found in a number of seeds germinating in light, such as *Lactuca sativa* and *Amaranthus retroflexus*. Prolonged far red irradiation induces a HIR and prevents germination in darkness in plants which usally

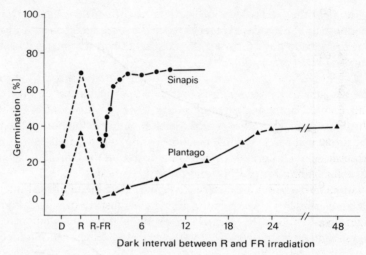

Fig. 10.22. The effect of dark period length on the disappearance of red/far red reversibility during germination (from Frankland).

germinate in the dark. The HIR of germination inhibition in light germinating plants differs from other HIR in that it counteracts the inductive photomorphosis. The photostimulation of germination is an inductive morphogenesis, the photoinhibition a HIR. The activity of the effector during the HIR of germination is still obscure. We only know that P_{fr} stimulates germination after a red pulse while $P_{fr}X$ inhibits germination.

In dicots germination can be either hypogeal or epigeal. In the first type the cotyledons stay below the soil surface and the epicotyl grows out of the soil, as in peas. In the latter case the cotyledons are raised above the surface by an elongation of the hypocotyl, as in bush beans; the epicotyl and the primary leaves grow afterwards. In both types the shoots form a hook shortly after germination so that the primary leaves or the cotyledons are protected while growing through the soil. As soon as the hook has emerged from the soil it opens under the influence of light; P_{fr} stimulates the hook opening.

10.4.2   Vegetative growth

After germination the growth of both monocot and dicot seedlings remains under light control. Young grass seedlings are characterized by a coleoptile which protects the growing primary leaves. In 1- or 2-day-old *Avena* seedlings, in which the cells have just started to elongate, white or red light induces growth in coleoptiles while it inhibits growth in older seedlings. In continuous white light the final coleoptile length is reached within 3 days. Dark-grown seedlings need a few days longer to reach the same height as light-grown ones, but continue to grow for some time so that they reach their final length after about 8 days (Fig. 10.23).

Coleoptiles and leaf growth are closely correlated in this early stage. After a few days the growth accelerates and the leaves break through the coleoptile and unfold. Unfolding of the primary leaves is also controlled by phytochrome. Some phytohormones such as gibberellins or cytokinins can substitute red light, but their action cannot be reversed by far red light. Abscisic acid, IAA and δ-aminolevulinate inhibit the light-induced unfolding as well as inhibitors of RNA and protein synthesis.

Some monocots (for example *Zea mays*) have an additional internodium, called a mesocotyl, which is the most light-sensitive organ we know. A red light dose of 10^{-2} J m^{-2}, which corresponds to the red component of 3 min moonlight, inhibits growth by 15%. The green safelight which is used in many experiments also inhibits mesocotyl growth. Inhibition increases by up to 50% of the dark control at red light doses of 10^{-1} to 10^2 J m^{-2} and is fully reversible below this value. Higher doses induce an inhibition of up to 95% of the dark control which is not reversible by far red; it is thus a typical HIR. Because of lacking reversibility either other P_{fr} forms or other phytochrome pools are responsible for these reactions than they are for inductive morphoses (see Section 10.2.4).

Hypocotyl growth in dicot seedlings basically depends on cell elongation, which is controlled by light (Fig. 10.24). Cell divisions are the basis for the longitudinal growth of hypocotyls and shoots. The growth curves for light- and dark-grown seedlings show a sigmoidal shape, i.e. the initial growth is

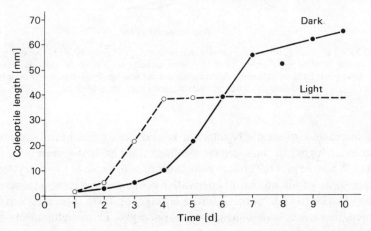

Fig. 10.23. Coleoptile growth in oat seedlings in darkness and continuous white light (from Thomson).

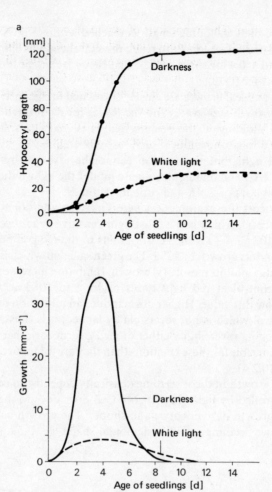

Fig. 10.24. (a) Hypocotyl growth curves of mustard seedlings (*Sinapis alba*). The seedlings were grown on a nutrient agar under exactly defined conditions. The only variable was light (continuous white light or darkness). Abscissa: days after sowing. (b) Hypocotyl growth rate of mustard seedlings (after data from Feger, from Mohr).

slow, increases exponentially afterwards and finally reaches a saturation, which is far higher in dark-grown seedlings than in light-grown ones (Fig. 10.24a). When we plot the growth rate, in terms of increase of hypocotyl length per time unit, we obtain the first derivative of the general growth curve (Fig. 10.24b). Dark-grown hypocotyls have a far higher growth rate and reach their final length earlier than light-grown hypocotyls. Thus, light affects both growth rate and duration.

The shoot growth of dicots and monocots is also controlled by the phytochrome system. The hypocotyls of etiolated and green seedlings react differently on red light. Continuous far red, red and blue light inhibit the hypocotyl growth in etiolated *Cucumis* or *Sinapis* seedlings (Fig. 10.25). In green seedlings far red either enhances growth, as in *Cucumis* or has no effect, as in *Sinapis*. In far red light etiolated and green *Lactuca* seedlings show behavior similar to that of cucumber seedlings, while red has no or only a small effect (see Fig. 10.15). It is assumed that etiolated seedlings contain largely labile phytochrome in two different forms, A and B (see Section 10.2.4).

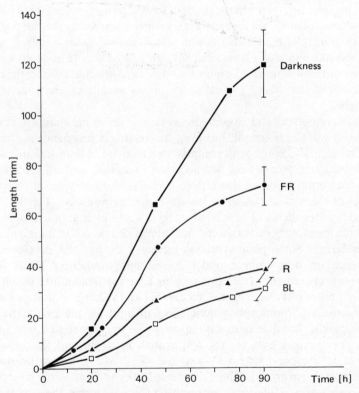

Fig. 10.25. Hypocotyl growth of etiolated cucumber seedlings in continuous far red, red and blue light (from Black and Shuttleworth).

It is assumed that photomorphogenetic reactions in green plants are regulated either by the P_{fr} concentration or the P_{fr}/P_{tot} ratio of the stable phytochrome pool. Results concerning the shoot growth of *Chenopodium* and other plants support the hypothesis that the photoequilibrium is the decisive factor. The shoot length of *Chenopodium* depends on the ratio between red and far red quanta of the irradiated white light in which they grow for a longer period of time. A high amount of red light (R:FR = 2.3) keeps the shoots short and a low one (R:FR = 0.2) induces them to grow longer. A high growth rate corresponds to a low P_{fr}/P_{tot} ratio (Fig. 10.26a). This dependence is especially important for plants growing in the shade of others.

Figure 10.26b demonstrates that in the range of 0 to 0.6 the photoequilibrium and thus–assuming a constant P_{tot}–the P_{fr} concentration also strongly depends on the red/far red ratio in the irradiated light. This ratio changes drastically during the day, especially in the twilight of dawn and dusk. For example, plants growing in the shade of trees receive far more far red light than non-shaded plants since the red light is mainly absorbed by the chlorophylls of the shading plants. Therefore sun plants grown under shade conditions respond like etiolated plants, and have elongated shoots, reduced leaf areas and fewer ramifications. The increased shoot growth could enable the plant to grow out of the shade. Typical shade plants usually cannot survive in full sunlight.

The development and growth of leaves is also under light control. The initial stages of leaf development, however, are relatively independent of external factors. Usually, leaf growth comprises an increase in both leaf area and leaf thickness, and depends on cell divisions as well as on cell growth. Light enhances both processes more than darkness, for example, in the primary leaves of *Phaseolous vulgaris* a few days after germination (Fig. 10.27). The palisade parenchyma of irradiated leaves 2 weeks after germination contains about twice as many cells as in etiolated plants and the cell volume is about 6–8 times larger. Since photosynthesis increases during leaf development its influence on development under continuous irradiation seems feasible. However, short light pulses or prolonged far red irradiation, which barely induce photosynthesis, increase cell number and volume, so that we also have to assume a photomorphogenetic effect of light for the first steps of leaf development. The fluence rate is important for the growth of leaves: in wheat high fluence rates enhance the leaf growth in width and thickness, while growth in area and length is inhibited (Fig. 10.28). Lower fluence rates enhance leaf area growth. Therefore shade leaves are thinner but larger in area than sun leaves. Differences in photosynthesis between the two types are discussed in Chapter 9.

10.4.3 Organelle development

The morphogenesis of a plant is closely linked to the organelle development. The effect of light on the morphogenesis of organelles has been studied in plastids, mitochondria and peroxisomes.

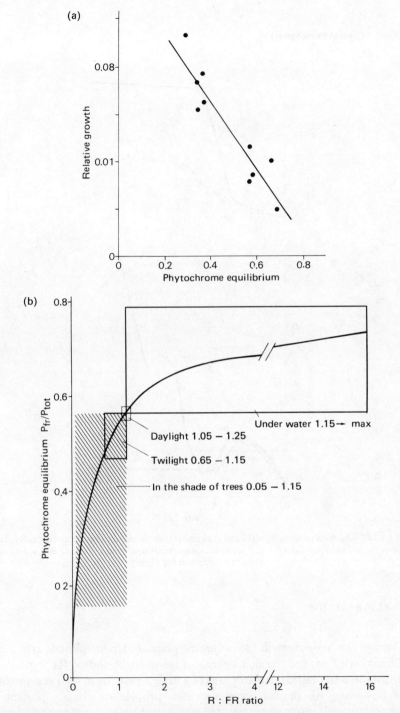

Fig. 10.26. (a) Relationship between the logarithm of shoot growth rate and the P_{fr}/P_{tot} ratio in *Chenopodium album* (from Morgan and Smith). (b) Relationship between the red/far red ratio of the light source and the phytochrome photoequilibrium (from Smith).

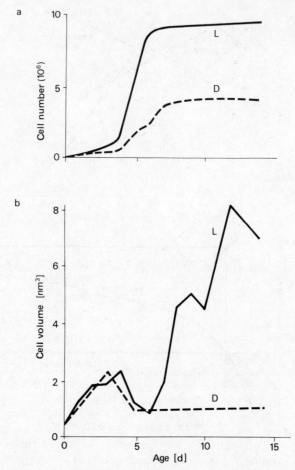

Fig. 10.27. The effect of white light (L) and darkness (D) on the cell number (a) and the cell volume (b) in the palisade parenchyma of *Phaseolus vulgaris* seedlings (from Gaba and Black, and data from Verbelen and De Greef).

10.4.3.1 PLASTIDS

During the development of a green plant different plastid types are differentiated in the various organs. Figure 10.29 shows the types and distribution of plastids in the organs of a higher plant from seed germination to flowering. All plastids originate from proplastids already present in meristemic cells, from where they are distributed to the daughter cells during cell division. Each plastid type has its own biogenesis, which is either light-independent as in etioplasts of dark-grown seedlings or light-dependent such

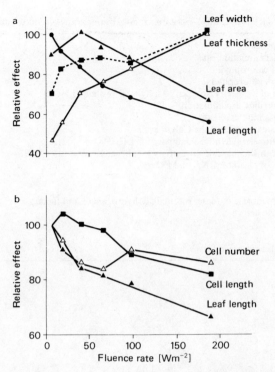

Fig. 10.28. Effect of the fluence rate on leaf development in wheat seedlings: (a) leaf length, area, width and thickness, (b) cell number, cell length and leaf length (shown as percentage of the values at the lowest fluence rate (7.6 W m^{-2}) (from Friend and Pomeroy).

as in chloroplasts of irradiated seedlings (Fig. 10.30). During irradiation etioplasts are transformed into chloroplasts. Though this is not the normal chloroplast development it allows us to draw valuable conclusions for the photomorphogenesis of chloroplasts.

Etioplasts are enclosed by a double membrane as all other plastids and contain a prolamellar body (PB) which often shows a crystalline array (Fig. 10.30). In light the prolamellar body disappears within a few hours and development of thylakoids commences. Development to a photosynthetically fully active chloroplast, in which the thylakoids are differentiated into grana and stroma thylakoids, is usually finished within 48–72 h. In addition to chlorophyll precursors, etioplasts contain many enzymes, structural proteins, lipids, redox systems and pigments necessary for photosynthesis. These components are accumulated during illumination and many reactions are phytochrome-controlled either by enzyme activation, such as some enzymes of the Calvin cycle, or by regulation of enzyme synthesis (Table 10.6).

Recently it has been demonstrated that phytochrome controls *de novo*

TABLE 10.6   SOME PHYTOCHROME-DEPENDENT, PHOTOMORPHOGENETIC REACTIONS IN ORGANELLES

Etioplast chloroplast
   chlorophyll accumulation rate
   production of chlorophyll *b*
   Shibata shift of chlorophyll *a*
   accumulation of light-harvesting chlorophyll *a/b* protein (LHCP)
   protochlorophyllide accumulation
   synthesis of 5-aminolevulinate
   activity of some enzymes of the Calvin cycle
   induction of translatable mRNA activity for LHCP
   induction of translatable mRNA activity for the small subunit of RuBP carboxylase
   accumulation of plastid mRNA and rRNA

Mitochondria
   activity of cytochrome oxidase, succinate dehydrogenase and fumarase
   change of substructure

Peroxisomes
   glycolate oxidase in green leaves

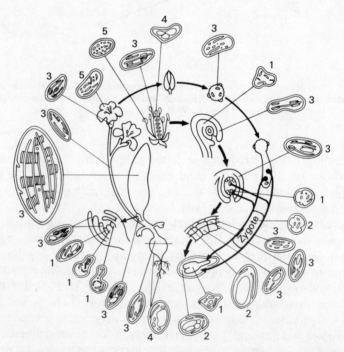

1 = Proplastid   2 = Amyloplast   3 = Chloroplast   4 = Leucoplast   5 = Chromoplast

Fig. 10.29. Diagram showing the various plastid types in the organs of higher plants during development. 1: Proplastid, 2: amyloplast, 3: chloroplast, 4: leukoplast, 5: chromoplast (from Leech).

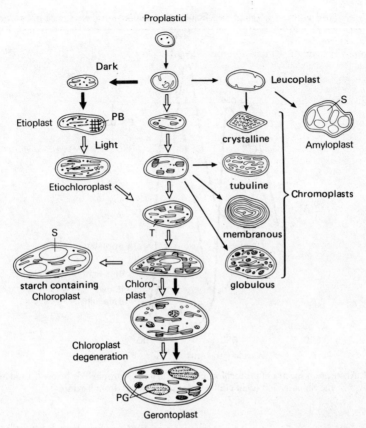

Proplastid

Dark

Leucoplast

Etioplast   PB

crystalline   Amyloplast

Light

tubuline   Chromoplasts

Etiochloroplast

membranous

T

starch containing
Chloroplast

Chloro-
plast

globulous

Chloroplast
degeneration

PG

Gerontoplast

Fig. 10.30. Schematic representation of the development of various plastid types.

syntheses on the level of mRNA, such as the small subunit of ribulosebisphosphate carboxylase. P_{fr} probably regulates the transcription rate of mRNA for this enzyme. Since this subunit is coded for in the nucleus, and synthesized in the cytoplasm, an exact coordination between nucleus, cytoplasm and stroma of the greening etioplasts is necessary. Phytochrome is not the primary photoreceptor for chlorophyll synthesis; this is protochlorophyllide, which is already present in etioplasts and is reduced to chlorophyllide in light. Geranylgeranyl phosphate is reduced to phytol, which is bound to chlorophyllide forming chlorophyll a which is then linked to the proteins in the primary thylakoids. These processes are concluded within 30 min and can be demonstrated in terms of spectral shifts from 652 nm to 678 nm and 684 nm and back to 672 nm (Fig. 10.31). The last absorption change is known as the Shibata shift, and is completed within 15 min in *Hordeum* seedlings. The Shibata shift is under phytochrome control since red light shortens the

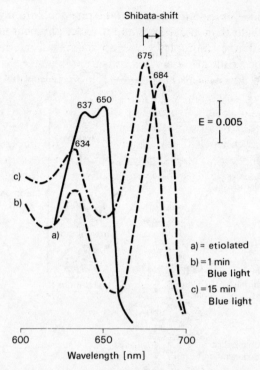

Fig. 10.31. Absorption spectra of protochlorophyllide (a) and chlorophyllide before (b) and after the Shibata shift (c) in etiolated *Hordeum* leaves (from Tevini).

duration of the shift considerably. Since chlorophyll is very photolabile during the Shibata shift a shorter period would be an advantage. Further developments are also enhanced by P_{fr} so that the process of greening is accelerated.

During this stage the synthesis of δ-aminolevulinate (ALA) is stimulated by phytochrome. In the dark the ALA synthesis is under feedback control of Mg-protoporphyrin. In light the limiting protochlorophyll pool is photoreduced to chlorophyllide and the ALA synthetase is controlled by phytochrome. The synthesis of chlorophyll for the light-harvesting chlorophyll–protein complex (LHCP) has also to be closely correlated with phytochrome-controlled protein synthesis. Since chlorophyll is synthesized in the stroma, but protein synthesis takes place in the cytoplasm, additional coordinating mechanisms are required. Phytochrome itself could regulate the transfer of transmitters by changing the permeability of the outer chloroplast membrane.

In roots and other underground organs amyloplasts are developed from proplastids. In light, these organs can also develop chloroplasts, but in contrast to leaf chloroplasts only blue light is effective. Gymnosperm seedlings also develop chloroplasts in the dark, but the rate of chlorophyll synthesis is much higher in light than in darkness and is under phytochrome control. In light-grown plants the chloroplasts develop via several intermediate forms. The primary thylakoids are generated by an unfolding of the inner plastid envelope; their synthesis is also under phytochrome control. Again information needs to be coordinated since part of the proteins are coded for in the nuclear DNA and part in the plastid DNA. An extreme example is the synthesis of RubP carboxylase: the small subunit is synthesized by the 80S cytoplasmic polysomes and the large subunit by the 70S chloroplast polysomes. Figure 10.32 summarizes the interactions and cooperation mechanisms between nucleus and plastid genome with the photoreceptors.

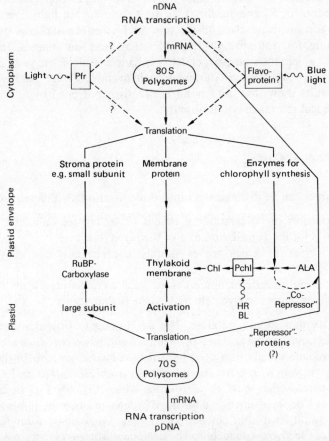

Fig. 10.32. Schematic diagram of the intracellular coordination between the genome of the nucleus (nDNA) and the chloroplasts (pDNA) as well as the influence of photoreceptors (from Parthier).

The substructure of chloroplasts changes with increasing age of the leaf. During the color change of leaves in autumn chloroplasts are transformed into gerontoplasts (senescent chloroplasts). During this development the thylakoids are degraded and plastoglobuli increase in volume (see Fig. 10.30). The transformation of chloroplasts into gerontoplasts takes place both in light and in darkness. Lack of minerals, drought or high temperatures can accelerate development.

10.4.3.2 MITOCHONDRIA

Little is known about the photomorphogenesis of mitochondria. Phytochrome effects on the ultrastructure and enzyme composition have been found in cotyledons of *Sinapis* seedlings. In white and far red light, typical plant mitochondria with tubular substructure are formed. In darkness the cristae type dominates, and strongly resembles that found in animals. In *Sinapis* seedlings a phytochrome-dependent enhancement of enzyme activities (cytochrome oxidase, succinate dehydrogenase and fumarase) has been found. These results show that mitochondria are also controlled by light, but the physiological relevance is still obscure.

10.4.3.3 PEROXISOMES

Peroxisomes can be distinguished into three functionally different types:

1. Peroxisomes in fat containing tissues of seeds predominantly contain enzymes for fat degradation and of the glyoxylate cycle.
2. Peroxisomes in green leaves mainly contain enzymes of glycolate metabolism.
3. Peroxisomes in heterotrophic tissues, such as in flowers, fruits or roots, contain enzymes for glycolate and ureate degradation.

All peroxisomes contain catalase. The development of peroxisomes in fat-containing reserve tissues is independent of light, while peroxisomes capable of the glycolate metabolism need light for their development. In dark (and light), cotyledons with fat reserves, such as from *Helianthus* and *Cucumis*, develop peroxisomes next to oleosomes which degrade fat. In light the cotyledons lose their function as storage organs and become photosynthetically active, at least after consumption of the fat reserves. Simultaneously enzymes of the glyoxylate metabolism disappear, such as isocitrate lyase, and the enzymes of the glycolate metabolism are synthesized. The pronounced increase of glycolate oxidase, the key enzyme of the glycolate metabolism, is initiated by phytochrome. The question whether synthesis and disappearance of enzymes are due to a metabolic change in the same peroxisomes, or whether new peroxisomes are formed, is still being discussed. Likewise it is unclear whether the enzymes are activated and deactivated or newly synthesized.

10.5 PHOTOMORPHOGENESIS IN ANIMALS

Light only controls the morphogenesis of animals in a few cases, such as the season dimorphism of butterflies. When caterpillars of *Araschnia levana* are kept in short days, brightly colored butterflies develop, and in long days, dark-colored. Thus, we find spring and summer forms with different wing patterns. The seasonal forms of dwarf cicadas differ not only in color and size but also in the anatomy of their reproductive organs. Light pulses interrupting the dark period influence the morphology of the seasonal forms. In contrast to plants, in which red light plays the major role in photomorphogenesis, in animals short-wavelength blue light and UV radiation are the most effective.

Blue light effects have been observed, especially in the circadian hatching rhythms of insects (see Chapter 11). In all cases the action spectrum resembles the absorption spectrum of flavins. Flavins are further supported by the fact that in insects both the wild type and a mutant, which lacks most of the carotenoids, show the same blue light-induced absorbance changes (LIAC). Furthermore, fruit flies, fed on a carotenoid-free diet, did not produce visual pigments, which depend on the availability of carotenoids (see Chapter 13), but they still showed the blue light-induced shift in the phase of their hatching rhythm.

10.6 BIBLIOGRAPHY

Textbooks and review articles

Björn, L. O. Photoreversibly photochromic pigments in organisms: Properties and role in biological light perception. *Quart. Rev. Biophys.* **12**, 1–23 (1979)

Briggs, W. R., Mandoli, D. F., Shinkle, J. R., Kaufman, L. S., Watson, J. C. and Thompson, W. F. Phytochrome regulation of plant development at the whole plant, physiological, and molecular levels. In: Colombetti, G., Lenci, F. and Song, P.-S. (eds.) *Sensory Perception and Transduction in Aneural Organisms*. Plenum Press, New York and London, pp. 265–280 (1985)

Dörnemann, D. and Senger, H. Blue-light photoreceptor. In: Smith, H. and Holmes, M. G. (eds.) *Techniques in Photomorphogenesis*. Academic Press, London, pp. 279–296 (1984)

Eilfield, P. and Rüdiger, W. Absorption spectra of phytochrome intermediates. *Z. Naturf.* **40c**, 109–114 (1985)

Ellis, R. J. Photoregulation of plant gene expression. *Bioscience Reports* **6**, 127–136 (1986)

Furuya, M. Molecular properties of phytochrome. *Phil. Trans. R. Soc. Lond.* **B,303**, 361–375 (1983)

Hartmann, K. M. A general hypothesis to interpret 'high energy phenomena' of photomorphogenesis on the basis of phytochrome. *Photochem. Photobiol.* **5**, 349–366 (1966)

Mancinelli, A. L. The high irradiance responses of plant photomorphogenesis. *Bot. Rev.* **44**, 129–179 (1978)

Marmé, D. Phytochrome: membranes as possible sites of primary action. *Ann. Rev. Plant Physiol* **28**, 173–198 (1977)

Mohr, H. Criteria for photoreceptor involvement. In: Smith, H. and Holmes, M. G. (eds) *Techniques in Photomorphogenesis*, Academic Press, London, pp. 13–42 (1984)

Mohr, H. and Sitte, P. *Molekulare Grundlagen der Entwicklung*. BLV, München (1971)

Mohr, H. and Schäfer, E. Photoperception and de-etiolation. *Phil. Trans. R. Soc. Lond. B*, **303**, 489–501 (1983)

Mohr, H. *Lectures on Photomorphogenesis*. Springer, Berlin, Heidelberg (1972)

Mohr, H., Drumm-Herrel, H. and Oelmüller, R. Coaction of phytochrome and blue/UV light photoreceptors. In: Senger, H. (ed.) *Blue Light Effects in Biological Systems*. Springer, Berlin, Heidelberg, New York and Tokyo, pp. 6–19 (1984)

Parthier, B. The role of phytohormones (cytokinins) in chloroplast development. *Biochem. Physiol. Pflanz.* **174**, 173–214 (1979)

Pratt, L. H. Phytochrome: the protein moiety. *Ann. Rev. Plant Physiol.* **33**, 557–582 (1982)

Pratt, L. H. Phytochrome purification. In: Smith, H. and Holmes, M. G. (eds) *Techniques in Photomorphogenesis*. Academic Press, London, pp. 175–200 (1984)

Quail, P. H. Phytochrome: a regulatory photoreceptor that controls the expression of its own gene. *TIBS* **9**, 450–453 (1984)

Quail, P. H., Colbert, J. T., Hershey, H. P. and Vierstra, R. D. Phytochrome: molecular properties and biogenesis. *Phil. Trans. R. Soc. Lond. B*, **303**, 387–402 (1983)

Roux, S. J. Phytochrome in membranes. In: Smith, H. and Holmes, M. G. (eds.) *Techniques in Photomorphogenesis*. Academic Press, London, pp. 257–277 (1984)

Rüdiger, W. Phytochrome, a light receptor of plant photomorphogenesis. *Struct. Bonding* **40**, 101–140 (1980)

Russo, V. E. A., Chambers, J. A. A., Degli–Innocenti, F. and Sommer, T. Photomorphogenesis in microorganisms. In: Colombetti, G., Lenci, F. and Song, P.-S. (eds) *Sensory Perception and Transduction in Aneural Organisms*. Plenum Press, New York and London, pp. 231–249 (1985)

Ruyters, G. Effects of blue light on enzymes. In: Senger, H. (ed.) *Blue Light Effects in Biological Systems*. Springer, Berlin, Heidelberg, New York and Tokyo, pp. 283–301 (1984)

Schäfer, E. A new approach to explain the "High Irradiance Responses" of photomorphogenesis on the basis of phytochrome. *J. Math. Biol.* **2**, 41–56 (1975)

Schäfer, E. Advances in photomorphogenesis. *Photochem. Photobiol.* **35**, 905–910 (1982)

Schäfer, E., Heim, B., Mösinger, E. and Otto, V. Action of phytochrome in light-grown plants. In: Vince–Prue, D., Thomas, B. and Cockshull, K. E. (eds) *Light and the Flowering Process*. Academic Press, London, pp. 17–32 (1984)

Schmidt, W. Physiological blue light reception. *Struct. Bonding* **41**, 1–44 (1980)

Schmidt, W. The study of basic photochemical and photophysical properties of membrane-bound flavins: The indispensible prerequisite for the elucidation of primary physiological blue light action. In: Senger, H. (ed.) *Blue Light Effects in Biological Systems*. Springer, Berlin, Heidelberg, New York and Tokyo, pp. 81–94 (1984)

Senger, H. The effect of blue light on plants and microorganisms. *Photochem. Photobiol.* **35**, 911–920 (1982)

Shropshire, W. Jr and Mohr, H. (eds): *Encyclopedia of Plant Physiology*, new series, Vols. 16A and 16B: *Photomorphogenesis*. Springer, Heidelberg (1983)

Smith, H. *Phytochrome and Photomorphogenesis*. McGraw-Hill, London, New York (1975)

Smith, H. (ed.) *Plants and the Daylight Spectrum*. Academic Press, London and New York, pp. 21–40 (1981)

Smith, H. Light quality, photoperception and plant strategy. *Ann. Rev. Plant Physiol.* **33**, 481–518 (1982)

Smith, H. and Grierson, D.: *The Molecular Biology of Plant Development*. Blackwell, Oxford (1982)

Smith, W. O. Jr.: Characterization of the photoreceptor protein, phytochrome. *Photochem. Photobiol.* **33**, 961–964 (1981)

Song, P. S.: Molecular aspects of photoreceptor function: phytochrome. In: Lenci, F. and Colombetti, G. (eds) *Photoreception and Sensory Transduction in Aneural Organisms*. Plenum Press, New York, pp. 235–240 (1980)

Further reading

Agnew, N., McCabe, A. and Smith, D. L. Photocontrol of spore germination in *Polypodium vulgare* L. *New Phytol*. **96**, 167–178 (1984)

Apel, K., Gollmer, I. and Batschauer, A. The light-dependent control of chloroplast development in barley (*Hordeum vulgare* L). *J. Cell Biochem*. **23**, 181–189 (1983)

Appenroth, K.-J., Hermann, G., Augsten, H. and Müller, E. Photokinetic and photophysical parameters of the blue light induced phototransformation of phytochrome in the presence of flavin. *Physiol. Plant*. **63**, 258–264 (1985)

Attridge, T. H., Black, M. and Gaba, V. Photocontrol of hypocotyl elongation in light-grown *Cucumis sativus* L. *Planta* **162**, 422–426 (1984)

Bartley, M. R. and Frankland, B. Phytochrome intermediates and action spectra for light perception by dry seeds. *Plant Physiol*. **74**, 601–604 (1984)

Beggs, C. J. and Wellmann, E. Analysis of light-controlled anthocyanin formation in coleoptiles of *Zea mays* L.: the role of UV-B, blue, red and far-red light. *Photochem. Photobiol*. **41**, 481–486 (1985)

Björn, L. O. Light-induced linear dichroism in photo–reversibly photochromic sensor pigments. V. Reinterpretation of the experiments *in vivo* action dichroism of phytochrome. *Physiol. Plant*. **60**, 369–372 (1984)

Blaauw-Jansen, B. Thoughts on the possible role of phytochrome destruction in phytochrome-controlled responses. *Plant Cell Environ*. **6**, 173–179 (1983)

Black, M. and Shuttleworth, J. E. The role of the cotyledons in the photocontrol of hypocotyl extension in *Cucumis sativus* L. *Planta* **117**, 57–66 (1974)

Borthwick, H. A., Hendricks, S. B., Toole, E. H. and Toole, V. K. Action of light on lettuce seed germination. *Bot. Gaz*. **115**, 205–225 (1954)

Braslavsky, S. E., Al-Ekabi, H., Petrier, C. and Schaffner, K. Phytochrome models: part 9. Conformation selectivity of the photocyclization of the biliverdin IX and IX dimethylesters. *Photochem. Photobiol*. **41**, 237–246 (1985)

Butler, W. L., Norris, K. H., Siegelman, H. W. and Hendricks, S. B. Detection, assay and preliminary purification of the pigment controlling photoresponsive development of plants. *Proc. Natl. Acad. Sci. USA* **45**, 1703–1708 (1959)

Bünning, E. and Mohr, H. Das Aktionsspektrum von Lichteinfluß auf die Keimung von Farnsporen. *Naturwissenschaften* **42**, 212 (1955)

Cordonnier, M.-M., Greppin, H. and Pratt, L. H. Characterization by enzyme-linked immunosorbent assay of monoclonal antibodies to *Pisum* and *Avena* phytochrome. *Plant Physiol*. **74**, 123–127 (1984)

Cosgrove, D. J. Rapid suppression of growth by blue light. Occurrence, time-course and general characteristics. *Plant Physiol*. **67**, 584–590 (1981)

Datta, N. and Roux, S. J. A rapid procedure for the purification of 124 kDalton phytochrome from *Avena*. *Photochem. Photobiol*. **41**, 229–232 (1985)

Dedonder, A., Rethy, R., Fredericq, H. and de Greef, J. A. Interaction between P_{fr} and growth substances in the germination of light-requiring *Kalanchoë* seeds. *Physiol. Plant*. **59**, 488–492 (1983)

Dring, M. J. and Lüning, K. Photomorphogenesis of marine macro-algae. In: Shropshire, W. Jr and Mohr, H. (eds) *Encyclopedia of Plant Physiology*. new series 16B: *Photomorphogenesis*. Springer, Berlin, pp. 545–568 (1983)

Drumm-Herrel, H. and Mohr, H. Mode of coaction of phytochrome and blue light photoreceptors in control of hypocotyl elongation. *Photochem. Photobiol*. **40**, 261–266 (1984)

Duke, S. O. and Lane, A. D. Phytochrome control of its own accumulation and leaf expansion in tentoxin- and norflurazon-treated mung bean seedlings. *Physiol. Plant*. **60**, 341–346 (1984)

Duysen, M., Eskins, K. and Dybas, L. Blue and white light effects on chloroplast development in a soybean mutant. *Photochem. Photobiol*. **41**, 667–672 (1985)

Eilfeld, P, and Rüdiger, W. On the reactivity of native phytochrome. *Z. Naturf*. **39c**, 742–745 (1984)

Ekelund, N. G. A., Sundqvist, C., Quail, P. H. and Vierstra, R. D. Chromophore rotation in 124-kDalton *Avena sativa* phytochrome as measured by light-induced changes in linear dichroism. *Photochem. Photobiol.* **41**, 221–223 (1985)

Ernst, D. and Oesterhelt, D. Purified phytochrome influences in vitro transcription in rye nuclei. *EMBO J.* **3**, 3075-3078 (1984)

Esashi, Y., Saitoh, H., Saijoh, Y., Ishida, S. and Kodama, H. Light actions in the germination of cocklebur seeds. II. Possible origin of the diversity of light responses in seed germination. *Plant Cell Physiol.* **26**, 361–370 (1985)

Flint, L. H. and McAlister, E. D. Wave lengths in the visible spectrum inhibiting the germination of light-sensitive lettuce seeds. *Smithson. Misc. Collect.* **96**, 1–8 (1937)

Fourcroy, P., Klein-Eude, D. and Lambert, C. Phytochrome control of gene expression in radish seedlings. II. Far-red light mediated appearance of the ribulose 1, 5-bisphosphate carboxylase and the mRNA for its small subunit. *Plant Sci. Lett.* **37**, 235–244 (1985)

Fourcroy, P., Lambert, C. and Klein-Eude, D. Phytochrome control of gene expression in radish seedlings. I. Far-red light mediated stimulation of polyribosome formation and appearance of translatable mRNAs. *Plant Sci. Lett.* **37**, 227–234 (1985)

Frankland, B. Germination in shade. In: Smith, H. (ed.) *Plants and Daylight Spectrum.* Academic Press, New York, pp. 187–204 (1981)

Friend, D. J. C. and Pomeroy, M. E. Changes in cell size and number associated with the effects of light intensity and temperature on the leaf morphology of wheat. *Can. J. Bot.* **48**, 85–90 (1970)

Fritz, B. J. and Ninnemann, H. Photoreactivation by triplet flavin and photoinactivation by singlet oxygen of *Neurospora crassa* nitrate reductase. *Photochem. Photobiol.* **41**, 39–45 (1985)

Gaba, V. and Black, M.: The control of cell growth by light. In: Shropshire, W. Jr and Mohr, H. (eds) *Encyclopedia of Plant Physiology*, new series 16A: *Photomorphogenesis.* Springer, Berlin, pp. 358–400 (1983)

Gaba, V., Black, M. and Attridge, T. H. Photocontrol of hypocotyl elongation in de-etiolated *Cucumis sativus* L. *Plant Physiol.* **74**, 897–900 (1984)

Hahn, T.-R., Chae, Q. and Song, P.-S. Molecular topography of intact phytochrome probed by hydrogen-tritium exchange measurement. *Biochem.* **23**, 1219–1224 (1984)

Hashimoto, T., Ito, S. and Yatsuhashi, H. Ultraviolet light-induced coiling and curvature of broom sorghum first internodes. *Physiol. Plant.* **61**, 1–7 (1984)

Heim, B., Jabben, M. and Schäfer, E. Phytochrome destruction in dark- and light-grown *Amaranthus caudatus* seedlings. *Photochem. Photobiol.* **34**, 89–93 (1981)

Heim, B. and Schäfer, E. Light-controlled inhibition of hypocotyl growth of *Sinapis alba* L. seedlings. Fluence rate dependence of hourly light pulses and continuous irradiation. *Planta* **154**, 150–155 (1982)

Hershey, H. P., Colbert, J. T., Lissemore, J. L., Barker, R. F. and Quail, P. H. Molecular cloning of cDNA for *Avena* phytochrome. *Proc. Natl. Acad. Sci.* **81**, 2332–2336 (1984)

Holmes, M. G. and Schäfer, E. Action spectra for changes in the 'high irradiance reaction' in hypocotyls of *Sinapis alba* L. *Planta* **153**, 267–272 (1981)

Holmes, M. G., Beggs, C. J., Jabben, M. and Schäfer, E. Hypocotyl growth in *Sinapis alba* L., the role of light quality and light quantity. *Plant Cell Environ.* **5**, 45–51 (1982)

Horwitz, B. A., Gressel, J., Malkin, S. and Epel, B. L. Modified cryptochrome *in vivo* absorption in dim photosporulation mutants of *Trichoderma. Proc. Natl. Acad. Sci.* **82**, 2736–2740 (1985)

Hsiao, K.-C. Sporulation in the fungus *Verticillium agaricinum:* Reversal of blue light inhibition by ultraviolet radiation. *Physiol. Plant.* **60**, 444–448 (1984)

Inoue, Y. and Furuya, M. Phototransformation of the red-light absorbing form to the far-red-light-absorbing form of phytochrome in pea epicotyl tissue measured by a multichannel transient spectrum analyser. *Plant Cell Physiol.* **26**, 813–819 (1985)

Inoue, Y. and Watanabe, M. Perithecial formation in *Gelasinospora reticulispora.* VII. Action spectra in UV region for the photoinduction and the photoinhibition of photoinductive effect brought by blue light. *Plant Cell Physiol.* **25**, 107–113 (1984)

Inoue, Y., Hamaguchi, H.-O., Yamamoto, K. T., Tasumi, M. and Furuya, M. Light induced fluorescence spectral changes in native phytochrome from *Secale cereale* L. at liquid nitrogen temperature. *Photochem. Photobiol.* **42**, 423–427 (1985)

Jabben, M. and Holmes, M. G. Phytochrome in light-grown plants. In: Shropshire, W. Jr and Mohr, H. (eds) *Encyclopedia of Plant Physiology*, new series 16B: *Photomorphogenesis.* Springer, Berlin, pp. 704–722 (1983)

Jabben, M., Beggs, C. J. and Schäfer, E. Dependence of P_{fr}/P_{tot} ratios on light quality and light quantity. *Photochem. Photobiol.* **35**, 709–712 (1982)

Jabben, M., Heihoff, K., Braslavsky, S. E. and Schaffner, K. Studies on phytochrome photoconversions *in vitro* with laser-induced optoacoustic spectroscopy. *Photochem. Photobiol.* **40**, 361–367 (1984)

Jenkins, G. I., Hartley, M. R. and Bennett, J. Photoregulation of chloroplast development: transcriptional, translational and post-translational controls? *Phil. Trans. R. Soc. B*, **303**, 419–431 (1983)

Johnson, C. B. The effect of red light in the high irradiance reaction of phytochrome: evidence for an interaction between P_{fr} and a phytochrome cycling driven process. *Plant Cell Environ. 3*, 45–51 (1980)

Johnson, C. B. and Tasker, R. A scheme to account quantatively for the action of phytochrome in etiolated and light-grown plants. *Plant Cell Environ.* **2**, 259–265 (1979)

Kasemir, H. Action of light on chlorophyll(ide) appearance. *Photochem. Photobiol.* **37**, 701–708 (1983)

Kazarinova-Fukshansky, N., Seyfried, M. and Schäfer, E. Distortion of action spectra in photomorphogenesis by light gradients within the plant tissue. *Photochem. Photobiol.* **41**, 689–702 (1985)

Kerscher, L. Subunit size, absorption spectra and dark reversion kinetics or rye phytochrome purified in the far-red absorbing form. *Plant Sci. Lett.* **32**, 133–138 (1983)

Lange, H., Shropshire, W., Jr. and Mohr, H. An analysis of phytochrome-mediated anthocyanin synthesis. *Plant Physiol.* **47**, 649–655 (1971)

Leech, R. M. The replication of plastids in higher plants. In: Mancinelli, A. L. and Schwartz, O. M. (eds) The photoregulation of anthocyanin synthesis. IX. The photosensitivity of the response in dark and light-grown tomato seedlings. *Plant Cell Physiol.* **25**, 93–105 (1984)

Mancinelli, A. L. and Walsh, L. Photocontrol of anthocyanin synthesis in young seedlings. *Plant Physiol.* **55**, 251–257 (1979)

Mandoli, D. F. and Briggs, W. R. Phytochrome control of two low-irradiance responses in etiolated oat seedlings. *Plant Physiol.* **67**, 733–739 (1981)

Mandoli, D. F. and Briggs, W. R. The photoreceptive sites and the function of tissue light piping in photomorphogenesis of etiolated oat seedlings. *Plant Cell Environ.* **5**, 137–145 (1982)

Mohr, H. and Oelze-Karow, H. Phytochrome action as a threshold phenomenon. In: Smith, H. (ed.) *Light and Plant Development*. Butterworth, London, pp. 257–284 (1976)

Mösinger, E. and Schäfer, E.: In-vivo phytochrome control of *in vitro* transcription rates in isolated nuclei from oat seedlings. *Planta* **161**, 444–450 (1984)

Mösinger, E. and Schopfer, P. Polysome assembly and RNA synthesis during phytochrome-mediated photomorphogenesis in mustard cotyledons. *Planta* **158**, 501–511 (1983)

Oelmüller, R. and Mohr, H. Induction versus modulation in phytochrome-regulated biochemical processes. *Planta* **161**, 165–171 (1984)

Oelmüller, R. and Mohr, H. Specific action of blue light on phytochrome-mediated enzyme syntheses in the shoot of milo (*Sorghum vulgare* Pers.). *Plant Cell Environ.* **8**, 27–31 (1985)

Otto, V., Mösinger, E., Sauter, M. and Schäfer, E. Phytochrome control of its own synthesis in *Sorghum vulgare* and *Avena sativa*. *Photochem. Photobiol.* **38**, 693–700 (1983)

Otto, V., Schäfer, E., Nagatani, A., Yamamoto, K. T. and Furuya, M. Phytochrome control of its own synthesis in *Pisum sativum*. *Plant Cell Physiol.* **25**, 1579–1584 (1984)

Quail, P. H., Colbert, J. T., Hershey, H. P. and Vierstra, R. D.: Phytochrome: Molecular properties and biogenesis. *Phil. Trans. R. Soc. Lond. B*, **303**, 387–402 (1983)

Revuelta, J. L. and Eslava, A. P. Photoregulation of carotenogenesis in *Phycomyces*. *Curr. Gen.* **8**, 261–264 (1984)

Roth-Bejerano, N. Growth control by phytochrome of de-etiolated bean hypocotyls mediated by chlorophyll fluorescence. *Physiol. Plant.* **50**, 326–330 (1980)

Rüdiger, W.: Chemistry of the phytochrome photoconversions. *Phil. Trans. R. Soc. Lond, B*, **303**, 377–386 (1983)

Rüdiger, W. and Thümmler, F. Low temperature spectroscopy of phytochrome: P_r, P_{fr} and intermediates. *Physiol. Plant.* **60**, 383–388 (1984)

Rüdiger, W., Thümmler, F., Cmiel, E. and Schneider, S. Chromophore structure of the physiologically active form (P_{fr}) of phytochrome. *Proc. Natl. Acad. Sci.* **80**, 6244–6248 (1983)

Ruzsicska, B. P., Braslavsky, S. E. and Schaffner, K. The kinetics of the early stages of the phytochrome phototransformation $P_r \rightarrow P_{fr}$. A comparative study of small (60 kDalton) and native (124 kDalton) phytochromes from oat. *Photochem. Photobiol.* **41**, 681–688 (1985)

Sarkar, H. K. and Song, P. S. Phototransformation and dark reversion of phytochrome in deuterium oxide. *Biochemistry* **20**, 4315–4320 (1981)

Schaer, J. A., Mandoli, D. F. and Briggs, W. R. Phytochrome-mediated cellular photomorphogenesis. *Plant Physiol.* **72**, 706–712 (1983)

Schäfer, E. and Mohr, H. Phytochromgesteuerte Photoantworten (photoresponses) im Dauerlicht und nach Lichtpulsinduktion. *Ber. Dtsch. Bot. Ges.* **98**, 85–98 (1985)

Schäfer, E., Beggs, C. J., Fukshansky, L., Holmes, M. G. and Jabben, M. A comparative study of the responsivity of *Sinapis alba* L. seedlings to pulsed and continuous irradiation. *Planta* **153**, 258–261 (1981)

Schäfer, E., Ebert, C. and Schweitzer, M.: Control of hypocotyl growth in mustard seedlings after light-dark transitions. *Photochem. Photobiol.* **39**, 95–100 (1984)

Schäfer, E., Lassig, T. U. and Schopfer, P. Phytochrome controlled extension growth of *Avena sativa* L. seedlings: fluence rate relationship and action spectra of mesocotyl and coleoptile responses. *Planta* **154**, 231–240 (1982)

Scheuerlein, R. and Braslavsky, S. E. Induction of seed germination in *Lactuca sativa* L. by nanosecond dye laser flashes. *Photochem. Photobiol.* **42**, 173–178 (1985)

Schmidt, W. Red/far-red photoreversibility: Not an appropriate phytochrome assay for red-light preirradiated corn coleoptiles. *Photochem. Photobiol.* **39**, 267–269 (1984)

Seyfried, M. and Schäfer, E. Phytochrome macro-distribution, local photoconversion and internal photon fluence rate for *Cucurbita pepo* L. cotyledons. *Photochem. Photobiol.* **42**, 309–318 (1985)

Seyfried, M. and Schäfer, E. Action spectra of phytochrome in vivo. *Photochem. Photobiol.* **42**, 319–326 (1985)

Sharma, R. Phytochrome regulation of enzyme activity in higher plants. *Photochem. Photobiol.* **41**, 747–755 (1985)

Smith, H. Phytochrome and photomorphogenesis in plants. *Nature* **227**, 665–668 (1970)

Spruit, C. J. P. Phytochrome intermediates in vivo IV. Kinetics of P_{fr} emergence. *Photochem. Photobiol.* **35**, 117–121 (1982)

Steinitz, B., Schäfer, E., Drumm, H. and Mohr, H. Correlation between far-red absorbing phytochrome and response in phytochrome-mediated anthocyanin synthesis. *Plant Cell Environ.* **2**, 159–163 (1979)

Sugai, M. and Furuya, M. Action spectrum in ultraviolet and blue light region for the inhibition of red-light-induced spore germination in *Adiantum capillus-veneris* L. *Plant Cell Physiol.* **26**, 953–956 (1985)

Sundqvist, C. and Björn, L. O Light-induced linear dichroism in photoreversibly photochromic sensor pigments. III. Chromophore rotation estimated by polarized light reversal of dichroism. *Physiol. Plant.* **59**, 263–269 (1983).

Sundqvist, C. and Widell, S. Binding of phytochrome to plasma membranes in vitro. *Physiol. Plant.* **59**, 35–41 (1983)

Tanada, T. Interactions of green or red light with blue light on the dark closure of *Albizzia* pinnules. *Physiol. Plant.* **61**, 35–37 (1984)

Tevini, M. Light, function, and lipids during plastid development. In: Tevini, M. and Lichtenthaler, H. K. (eds) *Lipids and Lipid Polymers in Higher Plants*. Springer, Berlin, Heidelberg and New York, pp. 121–145 (1977)

Thomas, B., Penn, S. E., Butcher, G. W. and Galfre, G. Discrimination between the red- and far-red-absorbing forms of phytochrome from *Avena sativa* L. by monoclonal antibodies. *Planta* **160**, 328–384 (1984)

Thompson, W. F., Everett, M., Polans, N. O., Jorgensen, R. A. and Palmer, J. D. Phytochrome control of RNA levels in developing pea and mung-bean leaves. *Planta* **158**, 487–500 (1983)

Thomson, B. F. The effect of light on cell division and cell elongation in seedlings of oats and peas. *Am. J. Bot.* **41**, 326–332 (1954)

Thümmler, F. and Rüdiger, W. Chromophore structure in phytochrome intermediates and bleached forms of phytochrome. *Physiol. Plant.* **60**, 378–382 (1984)

Thümmler, F., Eilfeld, P., Rüdiger, W., Moon, D.-K. and Song, P.-S. On the chemical reactivity of the phytochrome chromophore in the P_r and P_{fr} form. *Z. Naturf.* **40c**, 215–218 (1985)

Thümmler, F., Rüdiger, W., Cmiel, E. and Schneider, S. Chromopeptides from phytochrome and phycocyanin. NMR-studies of the P_{fr} and P_r chromophore of phytochrome and E, Z isomeric chromophores of phycocyanin. *Z. Naturf.* **38c**, 359–368 (1983)

Tobin, E. Phytochrome-mediated regulation of messenger RNAs for the small subunit of ribulose 1,5-bisphosphate carboxylase and the light-harvesting chlorophyll a/b-protein in *Lemna gibba*. *Plant. Mol. Biol.* **1**, 35–51 (1981)

Van der Woude, W. J. and Toole, V. K. Studies of the mechanism of enhancement of phytochrome-dependent lettuce seed germination by prechilling. *Plant Physiol.* **66**, 220–224 (1980)

Verbelen, J. P. and De Greef, J. A. Leaf development of *Phaseolus vulgaris* L. in light and darkness. *Am. J. Bot.* **66**, 970–976 (1979)

Vierstra, R. D. and Quail, P. H. Native phytochrome: Inhibition of proteolysis yields a homogeneous monomer of 124 kilodaltons from *Avena*. *Proc. Natl. Acad. Sci. USA* **79**, 5272–5276 (1982)

Vierstra, R. D. and Quail, P. H. Photochemistry of 124 kilodalton *Avena* phytochrome *in vitro*. *Plant. Physiol.* **72**, 264–267 (1983)

Vierstra, R. D. and Quail, P. H. Spectral characterization and proteolytic mapping of native 120-kilodalton phytochrome from *Cucurbita pepo* L. *Plant. Physiol.* **77**, 990–998 (1985)

Vierstra, R. D., Cordonnier, M.-M., Pratt, L. H. and Quail, P. H. Native phytochrome: immunoblot analysis of relative molecular mass and *in vitro* proteolytic degradation for several plant species. *Planta* **160**, 521–528 (1984)

Wada, M., Kadota, A. and Furuya, M. Intracellular localization and dichroic orientation of phytochrome in plasma membrane and/or ectoplasm of a centrifuged protonema of fern *Adiantum capillus-veneris* L. *Plant Cell Physiol.* **24**, 1441–1447 (1983)

Waddoups, D. R. Studies on variation in phytochrome-controlled seed germination in *Sinapsis arvensis*, Ph.D. thesis, University of London (1976)

Wall, J. K. and Johnson, C. B. An analysis of phytochrome action in the high-irradiance response. *Planta* **159**, 387–397 (1983)

Wendler, J., Holzwarth, A. R., Braslavsky, S. E. and Schaffner, K. Wavelength-resolved fluorescence decay and fluorescence quantum yield of large phytochrome from oat shoots. *Biochim. Biophys. Acta* **786**, 213–221 (1984)

Widell, S. and Sundqvist, C. Binding of phytochrome to plasma membrane *in vivo*. *Physiol. Plant.* **61**, 27–34 (1984)

Yatsuhashi, H. and Hashimoto, T. Multiplicative action of a UV-B photoreceptor and phytochrome in anthocyanin synthesis. *Photochem. Photobiol.* **41**, 673–680 (1985)

Zilberstein, A., Arzee, T. and Gressel, J. Early morphogenetic changes during phytochrome induced fern spore germination. I. The existence of a pre-photoinduction phase and the accumulation of chlorophyll. *Z. Pflanzenphysiol.* **114**, 97–107 (1984)

11 Light-dependent rhythms

11.1 INTRODUCTION

Many processes of life are rhythmic, such as the daily sleep of men and animals or the annual flowering and leaf fall of plants in the autumn. Rhythms which have a period of about 24 h are called circadian (*circa* = about, *dies* = day) and those which follow the year are called circannual (annuus = year).

Rhythmic leaf movements of Fabaceae such as *Tamarindus indica* or *Phaseolus vulgaris* were first observed by Androsthenes, a contemporary of Alexander the Great. During the day the leaves are raised and spread out, and during the night lowered and sometimes folded (Fig. 11.1). The time span between two maxima (or points of equal phase) of a rhythmic process is called the period and the distance between the highest and lowest leaf position is called the amplitude (Fig. 11.2). When synchronized by the natural day-and-night rhythm the period of leaf movements is 24 h. Under constant light or darkness the rhythmic leaf movements continue, which indicates that they are endogenous and independent of the external day and night rhythm. Under these conditions the period can be significantly different from 24 h. It is genetically determined and can range from 15 h to 30 h depending on the organism (Table 11.1).

Circadian and annual rhythms are common among plants and animals. In addition there are rhythms in the millisecond range, such as the activity of neurons, or in the second range, such as the heart beat or breathing. The circumnutations of the lateral leaflets of *Desmodium gyrans* have a period of several minutes. The terminal leaflets of this plant, however, follow a circadian rhythm. The rhythmic activities of some marine animals (*Carcinus*, *Mytilus*) and marine algae (*Fucus*) follow the tidal rhythm with a period of 12.4 h. Many marine organisms continue the tidal rhythm in an aquarium under constant conditions, which indicates a genetic determination. Some animals have a rhythm which follows the moon phases with a period of approximately 29.5 days.

Not all rhythmic processes are controlled by endogenous factors. Movement responses especially, such as photonastic and thermonastic movements, are often elicited exclusively by external factors. In other cases

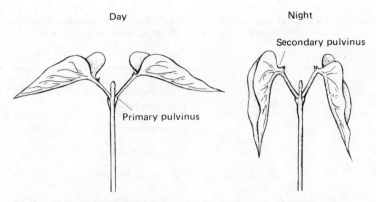

Fig. 11.1. Day and night position of *Phaseolus vulgaris* (from Bünning).

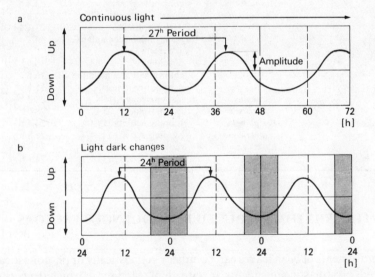

Fig. 11.2. Rhythmic leaf movements of *Phaseolus vulgaris*. (a) Circadian, free-running rhythm with a period of 27 h in continuous light; (b) rhythm synchronized by a light–dark regime with a period of 24 h (from Wagner, modified).

both the autonomous rhythm and external factors influence the response, as in the case of circumnutations of tendrils. Circumnutations can be described as swaying or circling movements of tendrils which aim to find an appropriate support. Both light and gravity are external factors which influence the response, and this makes it difficult to distinguish between endogenous and external control.

TABLE 11.1   EXAMPLES FOR CIRCADIAN RHYTHMS IN PLANTS AND ANIMALS PARTIALLY AT
DIFFERENT TEMPERATURES (AFTER BÜNNING, WAGNER MODIFIED).

Group of organisms	Organism	Reaction	Rhythm
Photosynthetic flagellates	*Gonyaulax polyedra*	luminescence photosynthesis	23–25 h
		rate, growth	22.5 h
		luminescence (15.9°C)	26.8 h
Algae	*Oedogonium cardiacum*	spore formation	22 h
Fungi	*Pilobolus sphaerosporus*	sporangial discharge	26.4 h
Ferns	*Selaginella serpens*	plastid morphology	*ca.* 24
Higher plants	*Phaseolus multiflorus*	leaf movements	27 h
	Kalanchoe blossfeldiana	petal movement	26 h
	Avena sativa	coleoptile growth	23.3 h
	Chenopodium	flower induction	30 h
		dark respiration	21–24 h
		betacyanine accumulation	24–30 h
		adenylate kinase activity	30 h
		chlorophyll accumulation	15 h
		net photosynthesis	15 h
	Phaseolus coccineus	leaf movement (15°C)	28.3 h
		leaf movement (25°C)	28.0 h
Animals	*Periplaneta americana*	running activity (19°C)	24.4 h
		running activity (25°C)	25.8 h
	Lacerta sicula	running activity (25°C)	24.3 h
		running activity (35°C)	24.1 h

11.2   RHYTHMS INDUCED BY EXOGENOUS FACTORS

The movements of bean leaves can be either nyctinastically or photonastically induced. Nyctinastic movements are circadian reactions in which the petioles lower in the primary pulvini at dawn while the leaf blades raise in the secondary pulvini (see Fig. 11.1). During dusk the movements are reversed. The mechanics is based on rhythmic, antagonistic changes in the turgor of the upper and lower halves of the pulvinus. In photonastic movements light causes a lowering of the leaf blade in the secondary pulvinus by decreasing the turgor pressure in both pulvinus halves. Even the decrease in the fluence rate caused by a cloud can induce a photonastic reaction.

Stomata also show a very sensitive photonastic reaction; the mechanics also depends on turgor changes. The petals of many plant species respond very fast to a decrease in the fluence rate by closing the flower (for example *Nuphar*,

Gentiana species and cacti). In many cases both photic and thermic stimuli are effective. Green leaves also show photonastic growth movement responses. Young leaves of *Impatiens* species respond to a sudden shading with an increase of the growth rate on the leaf surface. Older leaves show only turgor movements rather than growth changes.

The mechanisms of photonastic growth movements are largely unknown, but the involvement of the phytochrome system has been found in many cases, such as the leaflet movements of *Mimosa*.

11.3   ENDOGENOUS RHYTHMS

It can easily be tested whether a rhythm is endogenous by following the response under exactly controlled, constant conditions. Well-studied examples are the rhythmic leaf movements of many plant species, especially among the Fabales: at constant light and temperature beans show a free running endogenous rhythm with a period of 27–29 h. In the natural day-and-night cycle the free-running biological clock is synchronized to 24 h (see Fig. 11.2).

Rhythms with very short periods in the second or millisecond range also seem to be endogenous, since we do not find a timer ('Zeitgeber') in nature with such short periods.

Endogenous rhythms with a circadian period have been found in almost all eukaryotes but not in prokaryotes (Table 11.1). We can even find several circadian rhythms occurring simultaneously and independent of each other in an organism, such as mitosis frequency, photosynthesis and bioluminescence (see Chapter 7) in the dinoflagellate *Gonyaulax polyedra* (Fig. 11.3). In higher plants and animals, functions of either single organs or of the whole organism

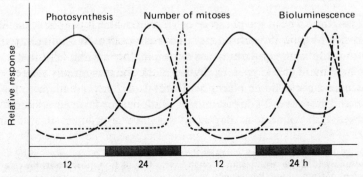

Fig. 11.3. Circadian rhythms of photosynthesis, mitosis frequency and bioluminescence of the marine flagellate *Gonyaulax polyedra* (from Hastings and Sweeney).

can be controlled by a circadian periodicity. It is still unknown how the individual rhythms in an organism are coordinated.

The deviation of the endogenous period from the natural day length can be extréme, such as the bean leaf movement with a difference of up to 4 h. Usually, deviations from the natural day are not longer than 30 min. In constant darkness, and at constant temperature, the endogenously controlled locomotory activity of flying squirrels is precisely maintained at 24 h and 21 min over a considerable period of time.

11.3.1   Synchronization of the endogenous rhythm

The examples in Table 11.1 show that free-running endogenous rhythms rarely have a period of exactly 24 h. But in the natural day circadian rhythms are synchronized to a period of 24 h. Light and temperature are the major factors responsible for synchronization.

In continuous darkness or light the circadian period of many animals and plants is largely independent of the ambient temperature (see Table 11.1). Neither the rhythmic activity of roaches or lizards nor the leaf movements of beans are markedly accelerated by increasing temperatures, which excludes the involvement of temperature-dependent metabolic processes in the endogenous rhythm. The Q_{10} values[1] of the period length usually vary between 1.0 and 1.1, and extreme values are 0.9 and 1.3: the physiological clock is temperature-independent. Oscillations with shorter periods, such as the movement of lateral leaflets of *Desmodium gyrans*, however, are shortened from about 5 min at 20°C to about 2 min at 30°C.

During the natural day-and-night cycle we find changes of both temperature and irradiance. Most plants and animals respond to the light/dark cycle rather than to the temperature cycle, as has been shown by experiments with a phase shift between the two signals. The light/dark change influences the period length as well as the phase of the rhythm, which is evident when plants are transferred from the natural day-and-night cycle into an inverse cycle in which they are artificially irradiated during night-time and kept in darkness during daytime. The endogenous rhythm of the plants is rapidly synchronized by the inverse cycle. Some plants need only 1 or 2 days to adapt to the new regime, and more than 14 days required for adaptation have never been found. In subsequent continuous darkness the inverse rhythm is maintained

---

[1]The Q_{10} value is a factor which indicates the enhancement of a reaction caused by a rise in temperature by 10°C. Physical reactions often have a Q_{10} of 1; in other words the reaction is independent of temperature. Biochemical reactions often have a Q_{10} of about 2 within certain physiological limits, which means that the reaction rate doubles with each temperature increase of 10°C.

(Fig. 11.4a). Light/dark regimes with periods shorter than 15 h or longer than 27 h are usually not accepted. The leaf movements of *Canavalia ensiformis* follow their endogenous rhythm when the plant is subjected to a 6:6 h light–dark regime (Fig. 11.4b). The endogenous rhythm can also be synchronized by a short light pulse during the dark period or a short modulation of the irradiance during the light period.

The circadian leaf movements of *Canavalia ensiformis* can be synchronized by regular short light pulses, as can many other plants and animals. In continuous darkness, interrupted each evening by a 2-h light pulse, the time of the lowest leaf position is shifted toward later times, so that the light pulse falls into the subjective day position of the leaves. When the light pulse is applied in the morning while the leaves rise, the time of the lowest position is advanced so that the light pulse again falls into the subjective day position. These phase shifts are only carried out, however, when the light pulses are not applied in too-short intervals.

Under natural conditions the changing light intensities during sunrise and sunset seem to be the most important factor for synchronization. In some cases

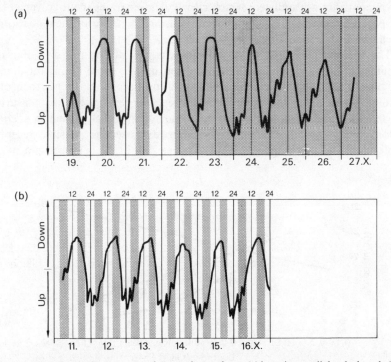

Fig. 11.4. Circadian leaf movements of *Canavalia ensiformis* (a) in an inverse light–dark cycle (left) and in continuous darkness (right) and (b) in a 6:6 hour light–dark cycle: no synchronization (from Kleinhonte).

experimentally produced changes in the illuminance of 1 lx induced a synchronization (Fig. 11.5). Changes in the illuminance of between 1 and 10 lx can be measured about 0.5 h before sunrise and after sunset.

The period length depends on the illuminance in a characteristic manner. Night-active animals, such as mice, respond to increasing illuminances with an increase in the period length. In day-active animals, such as chaffinches, the period length decreases with increasing illuminances (Fig. 11.5). The frequency of the rhythm changes linearly with the logarithm of the illuminance. This dependency (Aschoff's rule) has been shown to be valid both in mammals and birds. Plants also change their period length in response to the illuminance: in continuous light the period length of the bean leaf movements is about 24 h at 1 lx and about 21 h at 100 lx (Fig. 11.5).

In addition to the illuminance, the spectral composition is important for synchronization. In higher plants synchronization and changes in the period length are usually induced by red light. In dark-grown beans rhythmic leaf movements can be induced by transferring the plants into continuous red light. Far red radiation and light of other wavelengths are not effective. When far red light is administered simultaneously, or shortly after red light, leaf movements are not induced, which clearly demonstrates the involvement of the phytochrome system. Phytochrome is also responsible for the induction of circadian CO_2 production in *Bryophyllum*.

The spectral composition of daylight changes during the day. Far red dominates during dawn and dusk, and red light is more abundant during daytime. The change in red/far red irradiation influences the phytochrome system, which is thought to be the photoreceptor responsible for measuring daylength. In addition to the phytochrome system other pigments, such as chlorophylls, carotenoids and flavins, are discussed as possible receptor molecules. The action spectrum of the phase shift in the petal movements of

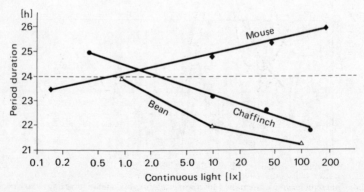

Fig. 11.5. Period length in dependence of the illuminance in mouse, chaffinch and bean (from Aschoff).

Kalanchoe blossfeldiana shows a maximum at 660 nm and a second one in the UV band at 366 nm. Neither the red nor the UV effect are reversible by far red. The unicellular alga *Gonyaulax* has two maxima in the action spectrum of the phase shift in the endogenous bioluminescence rhythm: at 475 nm and at 650 nm (Fig. 11.6). The comparison with the absorption spectrum indicates the activity of chlorophylls and possibly carotenoids. In many fungi and insects, blue light is effective inducing the phase shift of circadian rhythms (Fig. 11.6); the receptor is still unknown.

In animals blue and red light, as well as infrared radiation, synchronize endogenous rhythms. In most cases the light stimulus is perceived by the visual pigments; in arthropods and vertebrates extraretinal receptors located in the brain have been found. In birds the transparency of the skull is sufficient to allow illuminances of about 10 lx to reach the photoreceptor in the cerebrum, which is probably located in the pineal gland.

In most higher plants the leaves are the organs for light perception. The epidermis is especially sensitive to the light effective for synchronization; the upper epidermis is more sensitive than the lower one, as has been shown for *Kalanchoe*, *Glycine* and *Trifolium*.

The perception site is not always identical with the effector site, as has been shown for leaf movements in beans. When the lower leaves are kept in darkness and the upper leaves are exposed to a light/dark regime deviating from the circadian rhythm, all the leaves follow the induced rhythm, which indicates that there is information transduction between the leaves. The

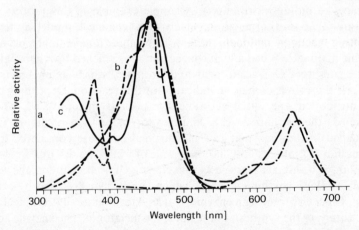

Fig. 11.6. Schematic action spectra for (a) the phase shift of petal movements in *Kalanchoe blossfeldiana* (after Schrempf, from Ninnemann); (b) phase shift in the hatching rhythm of *Drosophila pseudoobscura* (from Frank and Zimmermann); (c) conidia formation in *Neurospora crassa* (from Munoz and Butler); (d) phase shift in bioluminescence in *Gonyaulax polyedra* (from Hastings and Sweeney).

role of phytochrome in information transport has already been discussed in the chapter on photomorphogenesis (Chapter 10). The light sensitivity of the receptors changes both qualitatively and quantitatively during the day. These changes are not caused by changes in the receptor concentration or a degradation of the receptor during the day, as has long been assumed; they are rather due to periodic changes in the molecular milieu of the receptor. A possible reason for these fluctuations could be that at the beginning of the light period one or more substances are produced which accumulate like sand in a sand-glass, and are consumed gradually after the onset of darkness (sand-glass hypothesis). When the substance is consumed a new light response can commence. An alternative explanation is the circadian alteration of the molecular milieu by feedback: a substance produced and accumulated in light blocks its own synthesis until the concentration of the inhibiting substances is decreased due to metabolic degradation (Bünning hypothesis). A combination of both hyptheses is also feasible.

11.3.2   Molecular mechanisms of circadian rhythms

In contrast to the visible or measurable reactions, the basic molecular phenomena of induction and continuation of the circadian rhythms are largely unknown. It is assumed that there is not one single substance, the concentration of which oscillates between two extremes in a daily rhythm, acting as a zeitgeber. The biological oscillators are probably complicated sequences of biological reactions. A simple example of this process is the Belousov–Zhabotinski reaction, which is a chemical model system for oscillating reactions: malonate reacts with bromide and bromate, producing bromine malonate. When a high concentration is reached bromine malonate disintegrates to CO_2, acetic acid and bromide, which again reacts with malonate. Ferroin is used as an indicator for this reaction. Depending on the concentration of the initial substances (especially malonic acid, which is consumed) the oscillations can continue for more than an hour. Similar mechanisms on an enzymatic or biochemical basis are considered for the explanation of biological oscillations, some of which are discussed below. In both plant and animal tissues numerous enzymes have been found, the activity of which follows rhythmic changes which can be synchronized by external factors. Chicken produce an enzyme, N-acetyl transferase, in the pineal gland which catalyzes the synthesis of the hormone melatonin. The enzyme activity follows a distinctly circadian rhythm (Fig. 11.7). It is remarkable that the activity of the enzyme is inhibited by light.

It is, however, unlikely that such enzymatic rhythms are the molecular basis for circadian rhythms. The involvement of protein synthesis in biological

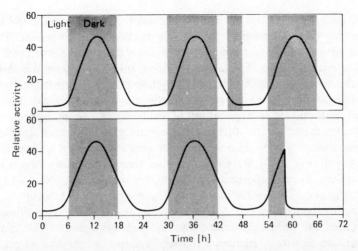

Fig. 11.7. Rhythmic activity of N-acetyl transferase in chicken in a light–dark regime. Above: additional dark time during the third light period: the rhythm does not change; below: premature ending of the third dark period: light inhibits the enzyme activity (from Binkley).

oscillations has been demonstrated by inhibition of several circadian rhythms with translation blockers. In some cases the involvement of the nucleus is also possible since a rhythmic DNA and RNA accumulation was found in liver cells. In *Acetabularia*, however, circadian rhythms have been observed to continue over several weeks under constant conditions after removal of the nucleus. After implantation of a nucleus from another plant the cell follows the rhythm of the new nucleus. Interactions between the nucleus, organelles and/or membranes are probably required for the generation of rhythms. This hypothesis is supported by the fact that circadian rhythms have not been demonstrated in prokaryotes.

In almost all models membrane processes play a predominant role; they are based either on changes in the membrane itself or on diffusion processes. The localization of the photoreceptors in membranes also indicates their involvement. Experimentally induced modifications in the fatty acid composition of the membrane in *Neurospora* (for example by temperature changes or by application of various membrane-active substances such as some steroids) induce a significant alteration in the period length of the circadian rhythm.

The ion composition of a cell also seems to be involved in the endogenous rhythm. Calcium ions seem to play the dominant role. Lithium ions – in competition with calcium and magnesium ions – inactivate ATPases, membrane receptors and several enzymes, and thus change the period length.

The only substance which always induces a lengthening of the period is

D_2O (heavy water). It is not known whether D_2O influences the rhythm by changing the ion transport through the membrane or by an exchange of H-atoms with D-atoms which could stabilize the protein structure.

The concentration of potassium which is involved in several membrane processes oscillates with a circadian rhythm. Pulses of increased potassium concentrations change the period length in several organisms. The same effect is found after application of valinomycin, an ionophore which increases the flux rates through the membrane. The importance of the membrane in the rhythmic transfer of ions is expressed in a model by Burgoyne (Fig. 11.8). Burgoyne assumes that monovalent ions, such as potassium, are transported through membrane proteins (Na-K-ATPase) into the cell. The synthesis of membrane proteins is blocked by high potassium concentrations. Since the membrane proteins are continuously degraded the potassium influx into the cell slows down. The resulting potassium deficiency induces a renewed synthesis of membrane proteins. This induces an increased potassium influx and the process continues. The lifetime of the membrane proteins, and the time needed for synthesis and transport, could explain a rhythm with circadian periods. Other authors assume similar processes at the chloroplast or mitochondria membrane but the *de novo* synthesis is replaced by a specific inhibition of membrane proteins.

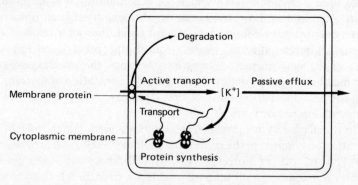

Fig. 11.8. Mechanism of the rhythmic regulation of K^+ ion transport through the membrane (from Burgyone, see text).

11.4 PHOTOPERIODISM

The control of biological processes by the day length has a number of ecological advantages. In addition to certain physiological functions the synchronization of reproduction is important for plants and animals. It is, for example, an advantage for animals to raise their offspring in favorable environmental conditions, and for plants not to flower in winter. Most plants and animals are obviously able to adapt their inner clock to the changing environmental conditions, especially to the changing length of the daylight period. Obviously they can measure the day or night length. The regulation of growth and environmental processes by a certain day length, the photoperiod, is called photoperiodism. The best-known example is the photoperiodic regulation of flowering. In addition to reproduction, senescence and dormancy phases as well as germination are controlled by the photoperiod.

11.4.1   Photoperiodism of flowering

In many plants flowering is regulated by a critical day length. There are three principally different types with different requirements for the day length:

1. *Short-day plants (SDP).* SDP (such as *Chenopodium rubrum*, *Kalanchoe blossfeldiana* and *Euphorbia pulcherrima*) flower when the day length is below a certain species-specific, critical value. Above this value the plants remain vegetative. In *Xanthium strumarium* (Fig. 11.9) the critical day length is 15.75 h, in other plants it is usually less.
2. *Long-day plants (LDP).* LDP (such as *Raphanus sativus*, *Sinapis alba* and *Spinacia oleracea*) flower when the day length exceeds a critical species-specific value. In *Hyoscyamus niger* (Fig. 11.9) this is more than 12.5 h, and in other plants more or less than that. In days shorter than the critical day length LDP remain vegetative and often form rosettes.
3. *Day-neutral plants.* In these plants (for example *Phaseolus vulgaris*, *Solanum tuberosum* and *Zea mays* flowering is independent of the day length.

In addition to plants with an absolute requirement for short or long days there are quantitative LDP and SDP; short or long days stimulate flowering but are not essential, such as in the SDP *Chrysanthemum multiflorum* and *Helianthus annuus*. Quantitative LDP are *Hordeum vulgare* and *Nicotiana tabacum* cv. Havanna. Some plants need two subsequent photoperiods of different lengths to induce flowering, and are consequently called short–long day plants (SLDP) and long–short day plants (LSDP), respectively. The LSDP *Kalanchoe laxiflora* flowers during short days in autumn after previous long days in the summer, but not in the short days in spring. The SLDP *Trifolium repens* needs short days followed by long days in the summer to flower.

In addition to the day length, the temperature plays an important role for flower induction. A number of plants need low temperatures to induce flowering, such as the quantitative SDP *Allium cepa*. Cereals sown in autumn, such as winter rye (*Secale cereale*) need the low temperatures in the winter to

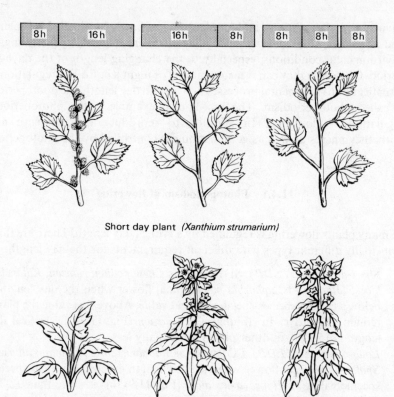

Short day plant *(Xanthium strumarium)*

Long day plant *(Hyoscyamus niger)*

Fig. 11.9. Flower induction in the short-day plant *Xanthium strumarium* and in the long-day plant *Hyoscyamus niger* under short-day (left) and long-day conditions (center) and with a light pulse during the dark phase (right) (from Raven).

flower in the long days of the following year. The effect of temperature on flower induction is called vernalization. Temperature also affects the critical day length. In the LDP *Hyoscyamus* the critical day length increases with increasing temperatures. *Euphorbia* is an extreme case, flowering in long days at low temperatures and in short days at high temperatures.

The fact that LDP flower above, and SDP below, a certain critical day length requires a precise measurement of the light period. According to Bünning's hypothesis photoperiodic plants have an endogenous oscillator which is synchronized by light/dark changes to 24 h. The oscillations of the endogenous oscillator consist of two half-phases, one of which is the photophile and the other the skotophile phase (Fig. 11.10). In SDP the photophile phase reaches its maximum after about 6 h (Fig. 11.10a) and in LDP only after about 16 h (Fig. 11.10d). The existence of such oscillations has

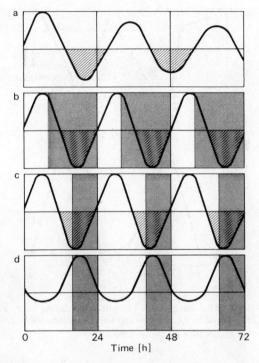

Fig. 11.10. Simplified hypothetical scheme for the measurement of the day length by an endogenous oscillator. (a) Circadian rhythm of the oscillator system with photophile (bright part of the curve) and scotophile (hatched part of the curve) phases; (b) short days with 8 h light period: the scotophile phase is always in the dark: SDP flower; (c) long days with 16 h light period, SDP do not flower; (d) long days with 16 h light period: the photophile phase receives light: LDP flower (modified from Bünning).

been demonstrated by interrupting the dark period with short light pulses. In *Kalanchoe blossfeldiana* a 1 h red light pulse in the middle of the dark period prevents flowering. The plants remain vegetative, although a short day preceded. LDP, such as *Fuchsia* cv. Lord Byron, which would remain vegetative in short days, are induced to optimal flowering by a 1 h light pulse given in the middle of the dark period (Fig. 11.11). The period of maximal sensitivity for a night break is different for different plant species and varieties but it occurs usually after 8–10 h darkness. This maximal sensitivity also follows a 24 h cycle, which indicates the existence of an endogenous, light-synchronized oscillator, the identity of which is still obscure. Judging from the results discussed above, light synchronizes the circadian rhythm in photoperiodism by interaction with an internal unknown oscillator. The result of this interaction is the prevention or promotion of flowering.

According to Bünning's model SDP measure the day length by testing

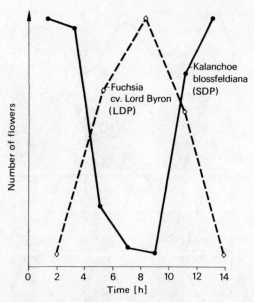

Fig. 11.11. Time kinetics of the sensitivity to 1 h light pulses during the dark phase in the flower induction of the LDP *Fuchsia* cv. Lord Byron and the SDP *Kalanchoe blossfeldiana* (from Vince-Prue).

whether light falls into the skotophile phase or not. This is the case when the days are shorter than the critical day length or the nights are longer than the critical night length. When the plant detects light in the skotophile phase flowering is suppressed.

In LDP flowering is prevented when the photophile phase is shortened. The light signal is perceived by the phytochrome system, as in many other photomorphogenetic processes. Action spectra for the night break show that both flower prevention in the SDP *Xanthium* and *Glycine*, and flower promotion in the LDP *Hordeum* and *Hyoscyamus*, have the same action maximum in the red wavelengths range and a minor peak in the blue range (Fig. 11.12).

The activity of red light pulses is different in SDP and LDP. Obviously LDP need a high amount of P_{fr} which is produced by a red light night break (Fig. 11.13). Flower induction in SDP, however, is inhibited by a high P_{fr} concentration. It is assumed that, at least in some SDP, active phytochrome needs to be degraded up to a certain threshold in order not to prevent flowering. This is the case when the dark phases (nights) are long enough. A red pulse restores the high P_{fr} concentration which, however, is only effective when the oscillator is sensitive to active phytochrome. This is the case in the middle of the dark phase. When red light is applied earlier or later, the inhibition of

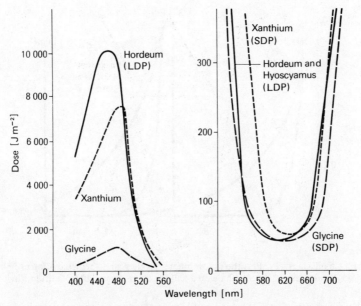

Fig. 11.12. Action spectra of the light effects during the dark phase in SDP and LDP (from Borthwick *et al.* and Parker *et al.*).

flower induction is less effective, as in the SDP *Kalanchoe blossfeldiana* (Fig. 11.13).

The light phase also shows an oscillating sensitivity, as can be demonstrated by red test pulses at different times. Long day and long night processes are phase-shifted in their phytochrome sensitivity (Fig. 11.14). The phytochrome-dependent flower induction suggests that synthesis and translocation of the proteins and enzymes necessary for promotion of flowering are also phytochrome-dependent.

According to a more recent model (Wagner, 1983) the phytochrome system regulates the translocation of proteins coupled to translation: P_{fr} reacts with a signal recognition protein (SRP) which corresponds to the receptor X' in Schäfer's model (see Fig. 10.18a). SRP can either be bound to a receptor site in the membrane or react with a signal peptide of a newly synthesized protein (Fig. 11.15). In the 'ribosomal' form it inhibits further synthesis of the polypeptide until the signal sequence finds the translocation-competent SRP site in the membrane. The polysome is also bound to the membrane by a receptor. When this happens the inhibition is released, protein synthesis continues and proteins are transported into the cell. The receptor sites for SRP, competent for translocation, are not always available, but depend on membrane properties and energy supply of the cell. In plant cells the energy

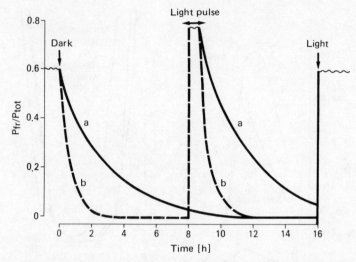

Fig. 11.13. Change in the $P_{fr}:P_{tot}$ ratio in continuous darkness and after irradiation of 30 min red light after 8 h darkness. P_{fr} destruction with a half-life of (a) 2 h and (b) 0.5 h (from Vince-Prue).

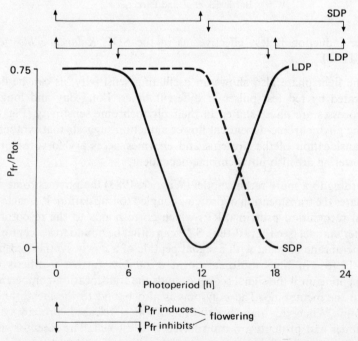

Fig. 11.14. Rhythmic sensitivity of an unknown oscillator on the active phytochrome form in LDP and SDP during a 24 h photoperiod (from Vince-Prue).

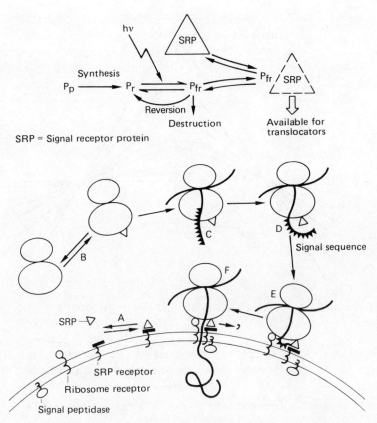

Fig. 11.15. Model for the effect of phytochrome on the synthesis and translocation of proteins involving the participation of ribosomes and membranes (from Wagner).

supply changes with the irradiation and can initiate circadian, endogenous rhythms of membrane functions. A reaction of active phytochrome P_{fr} with membranes or ribosomes is likely, since the fraction of phytochrome bound to cell structures changes rhythmically with periods of 24 h, 12 h and 6 h in darkness after a previous white light irradiation. In etiolated cucumber seedlings the absorption peak of P_{fr} could be shown to oscillate between 728 nm and 737 nm with a period of 22–24 h.

Many ecological problems are linked with photoperiodism, since day length depends on geographic latitude. Towards the poles the days are longer during summer and shorter during winter. Near the equator days and nights are always of almost equal duration (Fig. 11.16). During evolution many plant species have adapted their photoperiodic flowering to these different day lengths. Therefore near the equator we find almost exclusively SDP, while at latitudes >60° north or south we find predominantly LDP. Some soybean

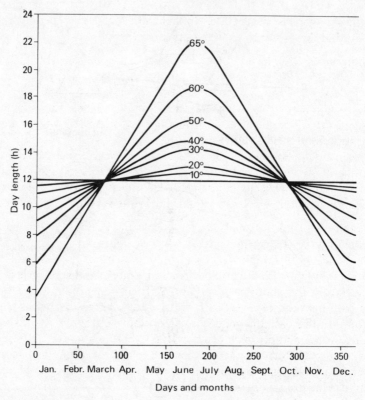

Fig. 11.16. Day length in dependence of the geographical latitude and the time of the year (from Salisbury and Ross).

varieties depend on a specific day length to such an extent that they grow optimally only within a zone of about 80 km width. This effect also indicates that genetically determined, endogenous rhythms control the band width for measurement of exogenous factors, especially light. It is largely unknown which primary signal decides when the measurement is initiated.

11.4.2   Photoperiodism in animals

In the whole animal kingdom we find a great variety of endogenous rhythmic growth, development and reproduction processes synchronized by light. In most vertebrates living outside the tropics the reproduction cycle is regulated by the photoperiod. Most birds are long-day animals; their gonads develop during the long days in early summer. In the short days of late summer and autumn the gonads degenerate. In hens the egg-laying period can be prolonged in the winter by extending the light period.

In migratory birds such as the garden warbler (*Sylvia borin*) light regulates the developmental phases which influence the induction of migration. Twice a year, in spring and in late summer, the body weight increases by additional fat reserves, which is closely correlated with the induction of migration. The molt, however, occurs at reduced body weight and not during migration periods. Changing the length of the light and dark phases experimentally cannot significantly alter the correlation between induction of migration and increase in body weight, which indicates that this correlation is genetically determined. The precise synchronization of the annual migration is, however, regulated by the seasonal photoperiod.

Measurement of the seasonal day length by birds and many mammals is explained by the coincidence model, which assumes that in animals, as in plants with photoperiodic reactions, photosensitivity is governed by a circadian rhythm. Photoperiodic reactions are stimulated when light coincides with the photosensitive periods. Experimentally testing the dark phase with light pulses reveals a rhythmic, endogenous photosensitivity in sparrows. In this case the gonadotropic hormone secretion was taken as a measure for the phototrophic reaction (Fig. 11.17).

The photoreceptors are the eyes and, in addition, extraretinal receptors such as the pineal organ in birds and dermal tissues in many invertebrates. In vertebrates the frontal organ and the parietal bodies are known as receptors, in addition to the pineal organ. In lizards it could be shown that the

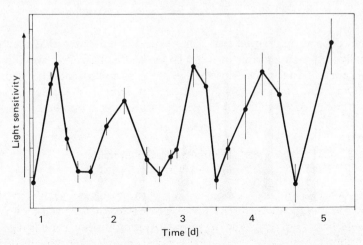

Fig. 11.17. Circadian rhythm of the photoperiodic sensitivity to light pulses during darkness in sparrows (from Follett, Mattocks and Farner).

locomotory and gonad activity, as well as the color changes, are exclusively controlled by the light sensitive pineal organ.

In insects such as *Drosophila melanogaster*, we find photoperiodic hatching rhythms similar to those mentioned above. Other examples of day length regulated morphogenesis include the spring and summer forms of *Araschnia levana*. Spring forms with bright wing patterns develop from caterpillars grown under short-day conditions (less than 14 h). Summer forms with dark wings develop in long days of more than 16 h. In dwarf cicadas there are seasonal forms which differ not only in their coloring but also in the morphology of their genital organs. As in some plants, we also find adaptations to a geographically determined day length in insects. Strains of *Drosophila litoralis* from northern Finland have a far longer critical day length (19 h), inducing their hibernating period, than strains from the Caucasus with a critical day length of only 12 h. In most insects the action spectra for photoperiodic reactions have a maximum in the blue range near 450 nm, which indicates carotenoids or flavins as possible photoreceptor pigments (see Chapter 10). In several other animal phyla, however, red light with wavelengths near 600 nm has also been found to be effective. In some cases, such as the nudibranch *Aplysia* and the snail *Hymnaea stagnalis*, heme proteins have been identified as photoreceptor pigments.

In addition to rhythms controlled by daylight, some developmental processes in animals, such as the hatching cycles of several insects, are controlled by moonlight. In *Clunio marinus* rhythmic hatching activity could be induced experimentally with only 0.5 lx, which corresponds to moonlight illuminance.

The mechanisms and transduction chains of induction and synchronization

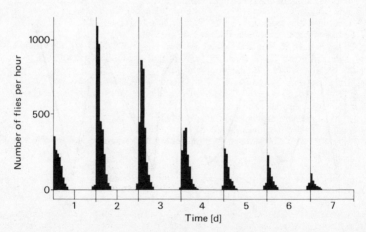

Fig. 11.18. Hatching rhythm in *Drosophila melanogaster* in darkness (starting on the second day) (from Pittendrigh and Bruce).

of photoperiodic reactions have been sufficiently revealed neither in the animal kingdom nor in the plant kingdom.

11.5 BIBLIOGRAPHY

Textbooks and review articles
Aschoff, J. *Circadian Clocks.* North Holland, Amsterdam (1965)
Aschoff, J. Phasenlage der Tagesperiodik in Abhängigkeit von Jahreszeit und Breitengrad. *Ökologia* **3**, 125–165 (1969)
Bernier, G. The factors controlling floral evocation: an overview. In: Vince-Prue, D., Thomas, B. and Cockshull, K. E. (eds) *Light and the Flowering Process.* Academic Press, London, pp. 277–292 (1984)
Brady, J. *Biological Timekeeping.* Society for Experimental Biology, Seminar Series 14 (1982)
Bünning, E. *Die physiologische Uhr.* Springer, Berlin (1977)
Deitzer, G. F. Photoperiodic induction in long-day plants. In: Vince-Prue, D., Thomas, B. and Cockshull, K. E. (eds) *Light and the Flowering Process.* Academic Press, London, pp. 51–63 (1984)
Engelmann, W. and Klemke, W. *Biorhythmen.* Biol. Arbeitsbücher No. 34, Quelle and Meyer, Heidelberg (1983)
Hamner, K. C. Photoperiodism and circadian rhythms. *Cold Spring Harbor Symp. Quant. Biol.* **25**, 269–277 (1960)
Haupt, W. *Bewegungsphysiologie der Pflanzen.* Thieme, Stuttgart (1977)
Haus, E. and Kabat, H. F. (eds) *Chronobiology* 1982–1983. Karger, Basel (1984)
Hillmann, W. S. Biological rhythms and physiological timing. *Ann. Rev. Plant Physiol.* **27**, 159–179 (1976)
Lang, A. Die photoperiodische Regulation von Förderung und Hemmung der Blütenbildung. *Ber. Dtsch. Bot. Ges.* **97**, 293–314 (1984)
Napp-Zinn, K. Light and vernalization. In: Vince-Prue, D., Thomas, B. and Cockshull, K. E. (eds) *Light and the Flowering Process.* Academic Press, London, pp. 75–88 (1984)
Ninnemann, H. Photoreceptors for circadian rhythms. In: Smith, K. C. (ed.) *Photochem. Photobiol. Rev.* 4, Plenum Press, New York (1979)
Porter, R. and Collins, G. M. (eds) *Photoperiodic Regulation of Insect and Molluscan Hormones.* CIBA Foundation Symposia. Vol. 104. Pitman, London (1984)
Rensing, L. Biorhythmik. *Unterricht Biol.* **4**, 1–46 (1980)
Rensing, L. and Schulz, R. Wie funktioniert die innere Uhr? *BIUZ* **14**, 13–19 (1984)
Salisbury, F. B. and Ross, C. W. *Plant Physiology.* Wadsworth, Belmont (1978)
Saunders, D. S. *Insect Clocks.* Pergamon, Oxford (1976)
Scheving, L. E., Halberg, F. and Pauli, J. E. *Chronobiology.* Thieme, Stuttgart (1974)
Schweiger, H. G. and Schweiger, M. Circadian rhythms in unicellular organisms: An endeavor to explain the molecular mechanism. *Int. Rev. Cytol.* **51**, 315–342 (1977)
Sweeney, B. M. *Rhythmic Phenomena in Plants.* Academic Press, New York (1969)
Vince-Prue, D. Light and the flowering process–setting the scene. In: Vince-Prue, D., Thomas, B. and Cockshull, K. E. (eds) *Light and the Flowering Process.* Academic Press, London, pp. 3–15 (1984)
Wagner, E. Biologische Uhr und Photoperiodismus. *BIUZ* **6**, 171–179 (1975)
Wagner, E. Molecular basis of physiological rhythms. In: Jennings, D. H. (ed.) *Integration of Activity in the Higher Plant.* Symposia of the Society of Experimental Biology, **31**, 33–72 (1977)
Wagner, E., Haertle, U., Kossmann, I. and Frosch, S. Metabolic and developmental adaptation of eukaryotic cells as related to endogenous and exogenous control of translocators between subcellular compartments. In: Schwemmler, W. and Schenk, H. (eds) *Endocytobiology*, II. Walter de Gruyter, Berlin and New York (1983)

Widell, S. and Larsson, C. Blue light effects and the role of membranes. In: Senger, H. (ed.) *Blue Light Effects in Biological Systems*. Springer, Berlin, Heidelberg, New York and Tokyo, pp. 177–184 (1984)

Winfree, A. T. *The Geometry of Biological Time*. Springer, Berlin (1980)

Further reading

Binkley, S. Ein zeitmessendes Enzym in der Zirbeldrüse. *Spektrum Wiss.* **6**, 82–89 (1979)

Bünning, E. Circadian rhythms, light and photoperiodism: A reevaluation. *Bot. Mag. Tokyo* **92**, 89–103 (1979)

Burgoyne, R. D. A model for the molecular basis of circadian rhythms involving ion-mediated translational control. *FEBS Lett.* **94**, 17–19 (1978)

Chay, T. R. and Cho, S. H. On exploring the basis for slow and fast oscillations in cellular systems. *Biophys. Chem.* **15**, 9–13 (1982)

Chen, Y. B., Lee, Y. and Satter, R. L. Chronobiology of aging in *Albizzia julibrissin*. II. An automated, computerized system for monitoring leaflet movement; the rhythm in constant darkness. *Plant Physiol.* **76**, 858–860 (1984)

Cleland, C. F. Biochemistry of induction – the immediate action of light. In: Vince-Prue, D., Thomas, B. and Cockshull, K. E. (eds) *Light and the Flowering Process*. Academie Press, London, pp. 123–135 (1984)

Cornelius, G. and Rensing, L. Can phase response curves of various treatments of circadian rhythms be explained by effects on protein synthesis and degradation? *BioSystems* **15**, 35–47 (1982)

Cremer-Bartels, G., Krause, K., Mitoskas, G. and Brodersen, D. Magnetic field of the earth as additional Zeitgeber for endogenous rhythms? *Naturwissenschaften* **71**, 567–574 (1984)

Eckhardt, D. and Engelmann, W. Involvement of plasmalemma ATPases in circadian rhythm of the succulent herb *Kalanchoe blossfeldiana (Crassulaceae)*. *Ind. J. Exp. Biol.* **22**, 189–194 (1984)

Engelmann, W. and Schrempf, W. Membrane models for circadian rhythms. *Photochem. Photobiol. Rev.* **5**, 49–86 (1980)

Follett, B. K., Mattocks, P. W. and Farner, D. S. Circadian function in the photoperiodic induction of gonadotropin secretion in the white-crowned sparrow. *Proc. Natl. Acad. Sci USA* **71**, 1666–1669 (1974)

Fukshansky, L. A quantitative study of timing in plant photoperiodism. *J. Theor. Biol.* **93**, 63–91 (1981)

Goss, R. J. Photoperiodic control of antler cycles in Deer. VI. Circannual rhythms on altered day lengths. *J. Expl. Zool.* **230**, 265–271 (1984)

Goto, K., Laval-Martin, D. L. and Edmunds, L. N., Jr. Biochemical modeling of an autonomously oscillatory circadian clock in *Euglena*. *Science* **228**, 1284–1288 (1985)

Hadley, P., Roberts, E. H., Summerfield, R. J. and Minchin, F. R. Effects of temperature and photoperiod on flowering in soya bean [*Glycine max* (L.) Merrill]: a quantitative model. *Ann. Bot.* **53**, 669–681 (1984)

Hasegawa, K. and Tanakadate, A. Circadian rhythm of locomotor behavior in a population of *Paramecium* multimicronucleatum: its characteristics as derived from circadian changes in the swimming speeds and the frequencies of avoiding response among individual cells. *Photochem. Photobiol.* **40**, 105–112 (1984)

Hastings, J. W. and Sweeney, B. W. The action spectrum for shifting the phase of the rhythm of luminescence in *Gonyaulax polyedra*. *J. Gen. Physiol.* **43**, 697–706 (1960)

Hasunuma, K. Circadian rhythm in *Neurospora* includes oscillation of cyclic 3′,5′-AMP level. *Proc. Japan Acad.* **60B**, 377–380 (1984)

Heath, O. V. S. Stomatal opening in darkness in the leaves of *Commelina communis*, attributed to an endogenous circadian rhythm: control of phase. *Proc. R. Soc. Lond. B*, **220**, 399–414 (1984)

Hoffmanns-Hohn, M., Martin, W. and Brinkmann, K. Multiple periodicities in the circadian system of unicellular algae. *Z. Naturf.* **39c**, 791–800 (1984)

Kinet, J.-M. and Sachs, R. M. Light and flower development. In: Vince-Prue, D., Thomas, B. and Cockshull, K. E. (eds) *Light and the Flowering Process*. Academic Press, London, pp. 211–225 (1984)

Klein, S. E. B., Binkley, S. and Mosher, K. Circadian phase of sparrows: control by light and dark. *Photochem. Photobiol.* **41**, 453–457 (1985)

Kleinhonte, A. Über die durch das Licht regulierten autonomen Bewegungen der *Canavalia*-Blätter. *Arch. Neerl. Sc. Ex. Nat.* **IIIB** (5), 1–110 (1929)

Kreuels, T., Joerres, R., Martin, W. and Brinkmann, K. System analysis of the circadian rhythm of *Euglena gracilis*. II: Masking effects and mutual interactions of light and temperature responses. *Z. Naturf.* **39c**, 801–811 (1984)

Lakin-Thomas, P. L. and Brody, S. Circadian rhythms in *Neurospora crassa:* interactions between clock mutations. *Genet.* **109**, 49–66 (1985)

Lecharny, A. and Wagner, E. Stem extension rate in light-grown plants. Evidence for an endogenous circadian rhythm in *Chenopodium rubrum*. *Physiol. Plant.* **60**, 437–443 (1984)

Lenk, R., Queiroz-Claret, C., Queiroz, O. and Greppin, H. Studies by NMR of in vitro spontaneous oscillations of activity in enzymatic extracts. *Chem. Phys. Lett.* **92**, 187–190 (1982)

Lercari, B. Role of phytochrome in photoperiodic regulation of bulbing and growth in the long day plant *Allium cepa*. *Physiol. Plant.* **60**, 433–436 (1984)

Lumsden, P. J. and Vince-Prue, D. The perception of dusk signals in photoperiodic time-measurement. *Physiol. Plant.* **60**, 427–432 (1984)

Martin, W., Jörres, R., Kreusel, T., Lork, W. and Brinkmann, K. Systemanalyse der circadianen Rhythmik von *Euglena gracilis:* Linearitäten und Nichtlinearitäten in der Reaktion auf Temperatursignale. *Ber. Dtsch. Bot. Ges.* **98**, 173–186 (1985)

Munoz, V. and Butler, W. L. Photoreceptor pigment for blue light in *Neurospora crassa*. *Plant. Physiol.* **55**, 421–426 (1975)

Nakashima, H. Is the fatty acid composition of phospholipids important for the function of the circadian clock in *Neurospora crassa? Plant Cell Physiol.* **24**, 1121–1127 (1983)

Nakashima, H. Calcium inhibits phase shifting of the circadian conidiation rhythm of *Neurospora crassa* by the calcium ionophore A23187. *Plant Physiol.* **74**, 268–271 (1984)

Ninomiya, S. Response of the rhythmic leaf movements of soybean (*Glycine max* L. Merr) to the light intensity of light/dark cycles. *Plant Cell Physiol.* **25**, 1451–1457 (1984)

Nirmal, K. S., Patra, H. K. and Mishra, D. Phytochrome regulation of biochemical and enzymatic changes during senescence of excised rice leaves. *Z. Pflanzenphysiol.* **113**, 95–103 (1984)

Oota, Y. Measurement of the critical nyctoperiod by *Lemna paucicostata* 6746 grown in continuous light. *Plant Cell Physiol.* **26**, 923–929 (1985)

Pittendrigh, C. S. Circadian rhythms and the circadian organization of living systems. *Cold Spring Harbor Symp. Quant. Biol.* **25**, 159–189 (1960)

Prasad, B. N. and Tewary, P. D. Circadian rhythmicity and photoperiodic responses in redheaded buntings (*Emberiza bruniceps*). *Period. Biol.* **86**, 17–22 (1984)

Queiroz-Claret, C., Girard, Y., Girard, B. and Queiroz, O. Spontaneous long-period oscillations in the catalytic capacity of enzymes in solution. *J. Interdiscipl. Cycle Res.* **16**, 1–9 (1985)

Rinnan, T., Johnsson, A. and Götestam, K. G. Imipramine affects circadian leaf movements in *Oxalis regnellii*. *Physiol. Plant* **62**, 153–156 (1984)

Samuelsson, G., Sweeney, B. M., Matlick, H. A. and Prezelin, B. B. Changes in photosystem II account for the circadian rhythm in photosynthesis in *Gonyaulax polyedra*. *Plant Physiol.* **73**, 329–331 (1983)

Schmid, R. and Koop, H.-U. Properties of the chloroplast movement during the circadian chloroplast migration in *Acetabularia mediterranea*. *Z. Pflanzenphys.* **112**, 351–357 (1983)

Sulzman, F. M., Ellman, D., Fuller, C. A., Moore-Ede, M. C. and Wassmer, G. *Neurospora* circadian rhythms in space: A reexamination of the endogenous-exogenous question. *Science* **225**, 232–234 (1984)

Swann, J. M. and Turek, F. W. Multiple circadian oscillators regulate the timing of behavioral and endocrine rhythms in female golden hamsters. *Science* **228**, 898–900 (1985)

Takahashi, J. S., DeCoursey, P. J., Bauman, L. and Menaker, M. Spectral sensitivity of a novel photoreceptive system mediating entrainment of mammalian circadian rhythms. *Nature* **308**, 186–188 (1984)

Vince-Prue, D. The perception of light-dark transitions. *Phil. Trans. Roy. Soc. Lond.* B **303**, 523–536 (1983)

Vince-Prue, D. Contrasting types of photoperiodic response in the control of dormancy. *Plant Cell Environ.* **7**, 507–513 (1984)

Volknandt, W. and Hardeland, R. Circadian rhythmicity of protein synthesis in the dinoflagellate, *Gonyaulax polyedra:* A biochemical and radioautographic investigation. *Comp. Biochem. Physiol.* **77B**, 493–500 (1984)

Wada, M., Miyazaki, A. and Fujii, T. On the mechanisms of diurnal vertical migration behavior of *Heterosigma akashiwo* (Raphidophyceae). *Plant Cell Physiol.* **26**, 431–436 (1985)

Welker, H. A., Schühle, A. and Vollrath, L. Infradian rhythms of serotonin and serotonin-N-acetyltransferase in the pineal gland of male rats. *J. Interdiscipl. Cycle Res.* **14**, 273–283 (1983)

Wellensiek, S. J. The effect of periodic dark interruptions in long photoperiods on floral induction in the long-day plants *Silene armeria* L., annual *Lunaria annua* L. and *Samolus parviflorus* Raf. *J. Plant Physiol.* **117**, 257–265 (1984)

Wirz-Justice, A., Wever, R. A. and Aschoff, J. Seasonality in freerunning circadian rhythms in man. *Naturwissenschaften* **71**, 316–319 (1984)

12 Light-dependent movement responses

Many microorganisms, higher plants and animals respond to light with movement responses. As in photomorphogenetic effects (see Chapter 11) light is not used as an energy source but as a sensory stimulus. (The exception is photokinesis, which is covered in Section 12.1.2.) Some organisms are extremely sensitive toward light and the responses require only very low intensities, such as phototropism in *Phycomyces* or phototaxis in *Dictyostelium*, which orients in light well below the perception threshold of the bright adapted human eye. This response requires a considerable signal amplification of several orders of magnitude. The basic mechanisms involved are currently being studied by a number of research groups.

According to the terminology in photomovement, light-dependent movement responses can be roughly subdivided into two categories, depending on whether we are dealing with motile microorganisms or with parts of sessile organisms, such as branches or leaves, which orient with respect to light. In addition there is a host of light-dependent responses at the intracellular level involving plasma and organelles.

12.1 MOTILE ORGANISMS

Motile microorganisms have developed three basically different light-dependent motor responses which are triggered by different qualities of light, such as light direction or quantum flux density.

1. Phototaxis is a directed movement of organisms with respect to the light direction. It is defined as positive (Fig. 12.1a) when the organisms move toward the light source and as negative (Fig. 12.1b) when they move away from it. Many flagellates show a pronounced positive phototaxis in low fluence rates, while higher fluence rates trigger the opposite response. Some organisms move at a certain angle with respect to the light source; when perpendicular to the light direction the orientation is called dia- or

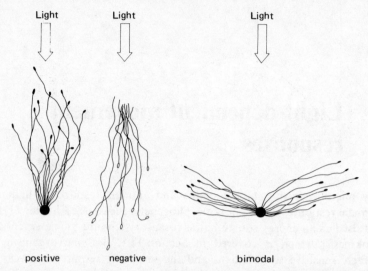

Fig. 12.1. Phototactic orientation of motile microorganisms with respect to light (arrows). (a) Positive phototaxis of pseudoplasmodia of the cellular slime mold *Dictyostelium discoideum*; (b) negative phototaxis of the flagellate *Euglena*; and (c) bimodal orientation of *Dictyostelium* pseudoplasmodia (from Poff, Fisher and Williams, Häder *et al.*, modified).

transversal phototaxis. Because of the lateral symmetry the distribution shows a bimodality (Fig. 12.1c).

2. Photokinesis describes the dependence of the speed of movement on the fluence rate. In contrast to phototaxis, this response is independent of the light direction. Photokinesis is defined as positive when an organism moves faster at a given fluence rate than in the dark control (Fig. 12.2). When it moves slower in light than in darkness, this is called negative photokinesis. Some organisms do not move in the dark, but are activated by light, while others are rendered immotile by light, especially at higher fluence rates (Lichtstarre). The dependence of the linear velocity of the organism on the fluence rate often follows an optimum curve where positive photokinesis occurs between a lower and upper limits (zero threshold and maximum).

3. Like photokinesis, photophobic responses are – unlike phototaxis – independent of light direction. But in contrast to both photokinesis and phototaxis, which are observed under steady-state conditions, photophobic responses are transient phenomena and are induced by sudden changes in fluence rate. Either a step-up or a step-down, or both, can cause phobic responses in a given species. Often step-down reactions are observed in low fluence rates while step-up reactions occur at higher fluence rates.

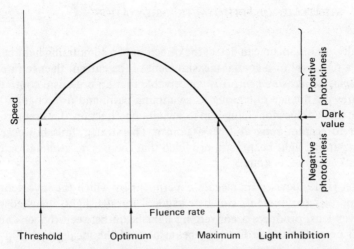

Fig. 12.2. Schematic diagram of the dependence of the speed of movement on the fluence rate (from Haupt, modified).

Thus the three basic light responses found in motile microorganisms differ with respect to the physical qualities of the stimulating light beam (Table 12.1). Nevertheless, it is sometimes difficult to pinpoint a given response to one of the types described above, since the phenomenological aspects have not been sufficiently revealed. Furthermore, several reactions often occur simultaneously in the same organism, and cannot always be separated experimentally.

TABLE 12.1   PHYSICAL QUALITIES OF STIMULI WHICH CAUSE LIGHT-DEPENDENT MOVEMENT RESPONSES IN MOTILE MICROORGANISMS.

Reaction	Physical quality	Symbol
Phototaxis	Light direction	$\tilde{I}$
Photokinesis	Fluence rate	I
Photophobic response	Change in fluence rate per unit time	dI/dt

12.1.1   Phototaxis

As early as 1803, Treviranus described the orientation of green flagellates in lateral light. Since then phototactic responses have been found in many photosynthetic and non-photosynthetic organisms in the animal and plant kingdom, among both prokaryotes and eukaryotes.

12.1.1.1 STRATEGIES OF PHOTOTACTIC ORIENTATION

Basically, an organism can detect the direction of a stimulating light beam by either a temporal or a spatial measurement of the current fluence rate.

1. *Temporal measurement*. It is conceivable that an organism continuously measures the fluence rate along its swimming path, and uses changes in the fluence rate as clues for changes in its swimming direction. This requires only one photoreceptor apparatus per organism. This strategy fails, however, in an ideally parallel light beam, and provided that there is no attenuation due to absorption in the medium.

Buder (1919) devised an elegant experiment by which the involvement of temporal measurements can be excluded. Parallel light, focussed with a convex lens, produces a converging light beam between the lens and the focal point (*F*) and a diverging beam beyond the focal point (Fig. 12.3).

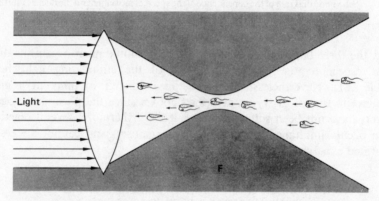

Fig. 12.3. Orientation of flagellates in a converging and diverging beam. When the organisms move through the focal point (F) toward the light source one can exclude an orientation based on a two-instant mechanism (see text, from Buder and Halldal, modified).

Organisms swimming in the diverging light beam toward the light source experience an increasing fluence rate. Some ciliates and flagellates continue to swim toward the light source even beyond the focal point, where they are necessarily subject to a decreasing fluence rate. Thus these organisms do not measure spatial changes in light intensity, but rather detect the direction of light with different mechanisms.

One possible mechanism has been discussed for those microorganisms which rotate while swimming forward along their longitudinal axis, such as many flagellates and ciliates. The fluence rate detected by the photoreceptor is modulated by the shading bodies or pigments which intercept the light path from the source to the receptor organelle. The unicellular photosynthetic flagellate *Euglena gracilis* swims helically by means of a single trailing flagellum (Fig. 12.4). In lateral light the stigma, consisting of lipid droplets colored deep orange by carotenoids and located dorsally next to the flagellar base, periodically casts a shadow onto the paraflagellar body (PFB, a swelling in the emerging flagellum) which is considered to be the photoreceptor in this organism. According to this hypothesis, the flagellum swings sideways, away from the cell body, each time the shadow of the stigma falls on the PFB. This turns the front end of the cell in the direction of the impinging light rays by a certain angle during each rotation, until it points toward the light source; then the stigma can no longer cast a shadow onto the PFB due to the geometrical properties of the cell.

This hypothetical mechanism may not be universal for all flagellates.

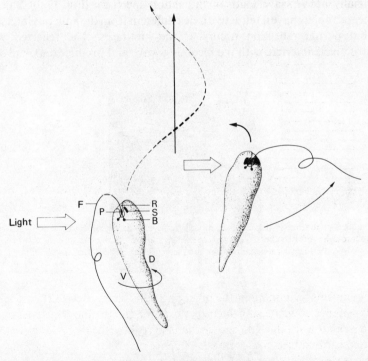

Fig. 12.4. Model of the phototactic orientation of *Euglena* according to the shading hypothesis. The cell moves on a helical track with its front end and ventral side (V) pointing outward. Both flagellae (F) originate from basal bodies (B) at the bottom of the reservoir (R), but one of them ends near the paraflagellar body (P) which is assumed to serve as the photoreceptor. The other is a trailing flagellum. In lateral light (arrows) the stigma (S) casts a shadow on the photoreceptor when the dorsal side (D) faces the light. This stimulus causes the flagellum to swing out so that the front end turns toward the light by a certain angle. This process is repeated during each revolution of the cell until the front pole points toward the light (from Checcucci and Haupt, modified).

Indeed, the Euglenophyta seem to be the exception, although they have been studied intensively. *Chlamydomonas*, a biflagellate green alga, also moves on a helical path. Its stigma, however, does not cast a shadow on the flagellar base since it is located on the equator of the almost spherical cell, slightly out of the plane of the flagellar bases. It has been suggested that the photoreceptor is located in the vicinity of the stigma. The urn-shaped chloroplast fills most of the cell body and shades the photoreceptor from the rear; therefore a course correction must be elicited by a periodic irradiation and not by a periodic shading (see above).

Recently, Foster and Smyth (1980) have proposed a novel explanation for light direction perception, based on reflection and interference in a quarter wavelength stack. In fact the stigmata of some *Chlamydomonas* species and several other flagellates have been found to consist of up to four layers of pigment globuli spaces at about 120 nm, which coincides with a quarter of the maximally active wavelength in the action spectrum (Fig. 12.5). The lipid pigment droplets have a refractive index different from the aqueous interstices, so that partial reflection occurs at the interfaces. The reflected waves constructively interfere with the incoming wave and produce maximal signal

Fig. 12.5. Electron micrograph of a section through the stigma of *Chlamydomonas reinhardtii* with four layers of pigment globules. Each layer is covered by a thylakoid double membrane from the inside. cm = cytoplasmic membrane, c = chloroplast (original photograph courtesy of L.A. Staehelin).

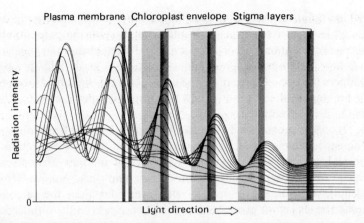

Fig. 12.6. Wave amplification by interference and reflection. Light strikes the cell surface perpendicularly from the left (highest curve) or tilted by 5° angles (until 70°, following curves). The wavelength is chosen to be 480 nm and the refractive index to be 1.5 for membranes, 1.6 for pigmented layers and 1.35 for nonpigmented layers (from Foster and Smyth, modified).

intensity near the presumed photoreceptor in the plasma membrane or the outer chloroplast membrane (Fig. 12.6). Indeed, the radiation reflected from the stigma can be detected microscopically. This model has a number of interesting consequences: first, the interference reflector is tuned to a certain wavelength range due to the thickness of the layers. Second, the signal detected by the photoreceptor has its maximum in a light beam perpendicular to the cell surface. During rotation the cell scans the horizon and uses the modulated signal for course correction. The modulation amplitude decreases while the cell turns toward the light direction, and is zero when it swims in the direction of the light source.

2. *Spatial measurement.* Organisms with more than one photoreceptor apparatus can measure internal light gradients and detect the light direction by comparison of the readings at different parts of the cell without moving. A typical example for this mechanism is the lens-shaped desmid, *Micrasterias* which, while standing on its edge, turns its face toward the light source.

A prerequisite for a spatial measurement is the existence of an internal light gradient. This can be produced by two antagonistic phenomena both in unicellular and in multicellular organisms.

a. When the irradiation is attenuated within the organism by absorption and/or scattering, the distal photoreceptors measure a smaller fluence rate than the proximal ones. This effect is being utilized by some cyanobacteria, desmids and amoebae of the cellular slime mold, *Dictyostelium*, as can be shown by partial irradiation of the organisms. These amoebae show positive phototaxis in low and negative phototaxis in high fluence rates. The direction of movement is defined by the development of pseudopodia at one end of the cell, which polarizes the cell at least temporarily. When the moving front is irradiated from above with a small intense light spot, the pseudopodia are withdrawn momentarily. Shortly afterwards the cell changes its polarity and extends new pseudopodia in a different direction. A weak light spot (several lux) induces the opposite reaction: the cells follow this light field when placed on the moving front and produce new pseudopodia when the side or the rear end of the cell is irradiated.

The second phenomenon is the lens effect, which has first been described for the phototropic bending of *Phycomyces* sporangiophores (see Section 12.2). Cylindrical or spherical organisms with a low internal absorption and scattering can focus parallel light onto their rear side, provided that the intracellular medium has a higher refractive index than the surrounding medium. The lens effect produces a higher fluence rate on the distal side than on the proximal side, which can be used as a clue for photoorientation. This mechanism is found in the phototaxis of *Dictyostelium* pseudoplasmodia (Fig. 12.7) and in the chloroplast reorientation in the filamentous green alga *Hormidium*. It may be a general rule that slow-moving, gliding organisms orient by means of intracellular spatial gradients, while rapid-swimming

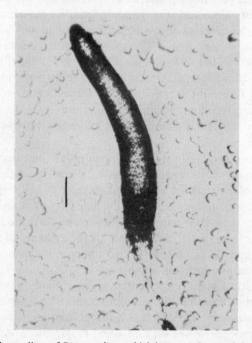

Fig. 12.7. Pseudoplasmodium of *Dictyostelium*, which has turned toward the lateral light source incident from the left. The light beam from below is focussed into a focal band on the top surface due to the lens effect in the pseudoplasmodium. The surrounding structures are nonaggregated amoebae. The scale represents 100 μm (from Häder and Burkart).

organisms (such as flagellates or ciliates) use temporal gradients, which can result in a signal modulation due to the rotation around the longitudinal axis.

The prokaryotic, photosynthetic cyanobacteria (cyanophyceae = blue–green algae) show a remarkable orientation in light. Filamentous species among the Hormogonales glide along their longitudinal axis when in contact with a substratum and rotate around this axis during forward movement (Fig. 12.8). The mechanical aspects of this motility have not yet been revealed; slime extrusion or contraction waves running over the surface are being discussed.

When a *Phormidium* population is inoculated in the center of an agar slab in a Petri dish, and irradiated laterally, it will spread out asymmetrically toward the light source. Microscopical observation shows, however, that the filaments do not turn actively toward the light. Front and rear ends of a trichome are morphologically identical, and the organisms reverse the direction of movement periodically every few minutes or hours. When they happen to move toward the light source reversals are suppressed for some time while they shorten the path away from the light source. Organisms more or less perpendicular to the light rays are not affected, but may change direction eventually when they hit an obstacle. Thus, over a certain period of time the population will move towards the light source. Phototactic orientation in gliding pennate diatoms is based upon similar behavior.

While *Phormidium* is a member of the Oscillatoriaceae, *Anabaena* belongs to

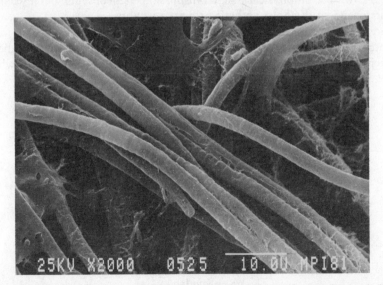

Fig. 12.8. Scanning electron micrograph of *Phormidium uncinatum* after cryofixation in liquid propane and gold sputtering. The constrictions mark the cross-walls. The fibrillar material is the remains of the slime sheaths. The scale represents 10 μm (original photograph courtesy of M. Claviez).

the family of the Nostocaceae. Unlike *Phormidium, Anabaena variabilis* is capable of a true course control with respect to the light direction. In lateral light the filaments turn toward the light source (or away from it in higher fluence rates, negative phototaxis). Partial irradiation experiments indicate that the organism compares the fluence rates on opposite sides of the cells.

12.1.1.2 PHOTORECEPTOR PIGMENTS

Different organisms have developed chemically very different pigment molecules as photoreceptors for radiation which control movement. Usually the photoreceptor molecules are identified experimentally by comparing action spectra with absorption spectra.

Phototactically orienting cyanobacteria use part of their photosynthetic pigments as photoreceptor molecules. Generally speaking, all photosynthetic prokaryotes seem to use the pigments of the energy-fixating machinery for photoorientation, as far as this has been studied up to now. However, the action spectra only coincide with the absorption spectra in certain wavelength regions. In the cyanobacterium, *Phormidium*, only the phycobilins, C-phycoerythrin and C-phycocyanin, as well as the carotenoids, seem to participate in photoorientation, while chlorophyll seems to be inactive (cf. Fig. 12.10). The biophysical signal transmission is still obscure, but it seems to be independent of photosynthetic primary reactions.

Some eukaryotic algae also seem to orient their movements in light using photosynthetic pigments. The action spectra of various desmids indicate activities in the red and blue regions of the spectrum (Fig. 12.9), which

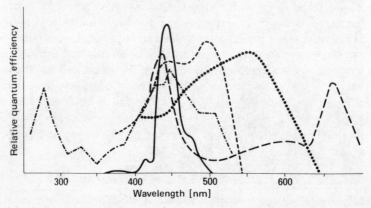

Fig. 12.9. Action spectra of the phototactic orientation of the flagellates *Gymnodinium* (dot–dashed), *Chlamydomonas* (short dashes), *Cryptomonas* (dotted), the gliding green alga *Cosmarium* (long dashes) and the red alga *Porphyridium* (continuous line) (from Nultsch, Watanabe and Furuya, Forward, Wenderoth and Häder, Nultsch and Schuchart, modified).

coincides well with the absorption spectrum. This indicates the participation of the chlorophylls. However, inhibitors of the photosynthetic electron transport chain, such as DCMU or DBMIB, do not impair phototactic orientation.

Within the red algae there are only a few (unicellular) motile species. Like cyanobacteria and cryptophyceae, red algae possess phycobilins as accessory pigments. The action spectrum, however, indicates that these pigments are not involved in phototaxis. In diatoms only the blue region is effective for photoorientation.

The different groups among the flagellates use different pigments, as shown by the action spectra. Some green flagellates and the unicellular *Euglena* seem to be an exception since they utilize a typical blue light receptor which is also used by fungi and higher plants for a variety of photomorphogenetic and behavioral responses. The molecule could be a membrane-bound flavin, but carotenoids are also discussed. The possible blue light photoreceptor pigments are described in Chapters 7 and 10, and in Section 12.2.1.1.

Many higher and lower animals also show a pronounced phototactic orientation. In most cases the photoreceptor is a rhodopsin or other retina pigments. Usually light direction detection is based on image perception or fluence rate measurements in more or less complex eyes.

12.1.2   Photokinesis

Photokinetic effects can easily be demonstrated by measuring the speed of an organism under different fluence rates. Another way of assaying photokinesis is by using a population technique: the dispersal area of a population from a central inoculation point is determined after a certain period of time. This technique integrates over the responses of a number of individual organisms.

In all photosynthetic prokaryotes studied so far, photokinesis seems to be linked to the energy-fixating process of photosynthesis. An increase in the fluence rate results in a higher production of ATP, which activates the motor apparatus. This is in accordance with the fact that the action spectra of purple bacteria and cyanobacteria resemble the absorption spectra of the photosynthetic pigments in these organisms (Fig. 12.10). Inhibition of the photophosphorylation by either uncouplers or inhibitors of the electron transport chain impair photokinesis.

12.1.3   Photophobic responses

Many microorganisms react with a photophobic response to a sudden change in the fluence rate. This response is independent of the stimulus strength as

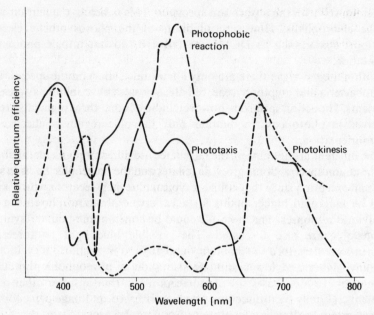

Fig. 12.10. Action spectra of phototaxis, photokinesis and photophobic reactions in *Phormidium uncinatum* (from Nultsch).

long as it exceeds a specific threshold. The phenomenological aspect of the response depends on the morphology of the organism. Purple bacteria and cyanobacteria reverse the direction of their movement while some flagellates and ciliates rotate on the spot or turn sideways. The term 'photophobic response' (Greek *phobos* = fright) has been coined by Engelmann (1882), who observed bacteria which reversed movement at a light/dark boundary, as if frightened.

Temporal as well as spatial changes in the fluence rate can induce photophobic responses. Some organisms reverse the direction of movement when they enter a light field projected into the suspension. This step-up response prohibits the organisms from getting into the light field, which is vacated within a short time. Organisms with a step-down response, however, are not hindered in entering the light field but reverse their direction of movement each time they try to leave the light field. Due to their random paths more and more organisms will enter the light trap and accumulate inside.

Since the intracellular signal amplification and transmission of a light stimulus is better known in cyanobacteria than in most other systems, we will discuss the sensory transduction pathway using this example. Formally, the reaction chain of events can be subdivided into three steps:

photoperception→sensory transduction→motor response

Photoperception describes the absorption of an effective quantum and the formation of an excited state of the photoreceptor molecule. The sensory transduction performs several functions: it transforms the signal from one form of energy into another, it amplifies the signal and transports the information inside the organisms. The last step is the control of the motor apparatus. It is important to stress that the absorbed light energy does not power the motor but rather controls its activity (e.g. induction or inhibition of motility or change of direction).

Both photosynthetic systems are involved in photoperception in the cyanobacterium *Phormidium*, as can be demonstrated by comparing the absorption spectrum with the action spectrum (Fig. 12.10). In addition, the (totally different) action spectra for photokinesis and phototaxis are shown to point out that these three light-dependent phenomena are independent of each other. (In other organisms two or more responses can be driven by the same photoreceptor.)

In contrast to photokinesis, which is linked to photophosphorylation, in all photosynthetic organisms studied in any detail photophobic responses are coupled directly to the photosynthetic electron transport chain. The coupling site seems to be plastoquinone, which is one of the redox systems involved in cyclic and non-cyclic electron transport and is located in the thylakoid membrane. In the prokaryotic cyanobacteria the thylakoid vesicles are not encapsulated in an additional plastid membrane, as in eukaryotes, and may be in contact with the cytoplasmic membrane (Fig. 12.11). In light, plastoquinone picks up protons from the cytoplasmic compartment and

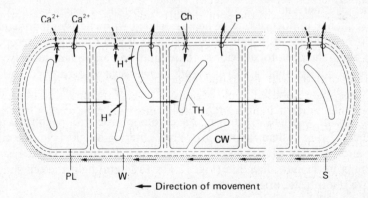

Fig. 12.11. Model for the sensory transduction of photophobic reactions in the cyanobacterium, *Phormidium*. The breakdown of the proton gradient across the thylakoid membrane (TH) in darkness induces an opening of voltage-dependent, selective ion channels (CH), which allow calcium ions to enter the cell through the cytoplasmic membrane (PL) along a gradient previously established by pumps (P). The arrows in the cross-walls (CW) symbolize a longitudinal current which returns in the slime sheath (S) and/or the cell wall (W), and builds up an electrical gradient along the trichome, and which dictates the direction of movement (from Häder, modified).

transports them into the interior of the thylakoid vesicles (cf. Section 9.1). The resulting pH gradient can amount to 2 to 3 pH units. In addition an electrical gradient $\Delta\Psi$ is generated; together these form the electrochemical potential $\tilde{\mu}_{H^+}$.

$$\tilde{\mu}_{H^+} = \Delta\Psi - \left(2.303\,\frac{RT}{F}\right)\Delta pH \text{ (in volts)}.$$

When the front cells of a trichome enter a dark zone the proton gradient probably breaks down very quickly since the protons diffuse out of the thylakoid vesicles back into the cytoplasm. This decreases the negative cytoplasmic electrical potential as measured compared to the exterior. In fact, changes in the electrical potential, due to a light/dark cycle, can be measured by inserting microglass electrodes into the cell. Recent results, however, show that these primary electrical potential changes open ion-specific, potential-dependent channels in the cytoplasmic membrane, which allows a massive influx of cations (presumably Ca^{2+}) along an ionic gradient previously established by active ion pumps. The influx of positive charges further depolarizes the cell, resulting in a considerable signal amplification.

The direction of movement of a trichome is dictated by an electrical gradient between front and rear end. Therefore this gradient has to be inverted each time the organism changes direction. It is still unknown how the potential gradient controls the motor apparatus (which is also obscure) and how the ionic current induced by darkening the front cells reverses the gradient.

Most ciliates are not colored, and respond to thermal, electrical, mechanical and chemical stimuli. Some ciliates also repond to light. *Paramecium* and a number of other ciliates show photophobic responses when exposed to a sudden change in fluence rate. When associated with endosymbiontic green algae (*Chorella*), they utilize the algal pigments as additional photoreceptors, which alters their photobehavior both qualitatively and quantitatively.

Colored ciliates such as *Stentor* and *Blepharisma* also respond to light. *Stentor* carries longitudinal rows of cilia along its pear-shaped body. Underneath the ciliar bases there are rows of vesicles which contain the photoreceptor pigment, stentorin. Its chromophoric group consists of a complicated ring system (hypericin). When entering a light field, the ciliate stops after a delay of 200 ms, turns sideways (30° to 180°) and resumes swimming in the new direction (Fig. 12.12). In lateral light the cell shows a negative phototactic orientation.

In no case the sensory transduction from the photoreceptor molecule to the motor apparatus has been fully analyzed. Recent results, however, indicate that in both flagellates and ciliates ionic currents and electrical potential changes play a major part. In the flagellate, *Haematococcus*, light-induced membrane potential changes could only be measured in the membrane portion overlaying the stigma. In *Chlamydomonas*, the direction of the

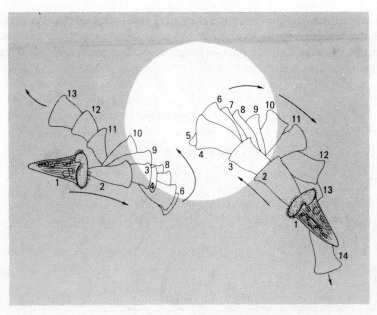

Fig. 12.12. Photophobic (step-up) reactions of the ciliate *Stentor* entering a light field. Its position is drawn every 1/6 s; The length of the organism is about 300 μm (from Song *et al.*).

flagellar beat can be influenced by electrical currents passed into the cell by microelectrodes. From this it would appear that electrical potential changes in these organisms are related to ionic currents, since the direction of the flagellar beat can also be controlled by calcium concentration.

Many other flagellated and non-flagellated, photosynthetic and non-photosynthetic microorganisms also show photophobic responses. In some cases we can speculate about the ecological advantages for an organism. Obviously it is advantageous for a photosynthetic organism to reverse the direction of movement each time it moves into a dark area. Likewise a step-up response is sensible for an organism whose pigments are photobleached in high fluence rates. In other cases the ecological significance of light-dependent movement responses is still obscure.

12.2   MOVEMENTS OF SESSILE ORGANISMS

Many sessile organisms respond to light with movement reactions. The movement of an organ with respect to light direction is defined as phototropism; bending toward the light source is called a positive tropism and away from it negative. Photonastic responses are also induced by light, but the direction is independent of light direction. Typical examples are the opening and closing of flowers, induced by sudden changes in fluence rate. The direction of movement is a built-in function of the organ and is only triggered by the stimulus (cf. Section 12.2.2).

12.2.1   Phototropism

Phototropic bending is usually based on a differential growth of opposite flanks of an organ. Shoots usually bend toward the light source, while roots are usually indifferent toward light and only a few cases of negative phototropism have been found. Leaves usually orient themselves perpendicular to the incident light (transversal phototropism). The classical definition of phototropism needs to be extended somewhat to include polarotropism, which is growth in a plane perpendicular to the electrical vector of linearly polarized light. This phenomenon has been found in germinating moss spores and growing pollen tubes (Fig. 12.13).

Though many organisms are known to show phototropism we will here

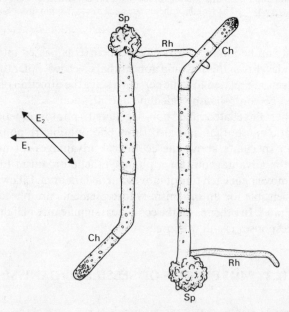

Fig. 12.13. Polarotropism in two germinating spores of *Dryopteris filix-mas*. The chloronema (Ch) grow perpendicularly, the rhizoids (Rh) parallel to the electrical vector E_1 of the polarized red light. After 9 days the polarization filter was turned by 50° (E_2) and the sample irradiated for another 8 h and left in darkness for an additional 16 h period (drawn after a microphotograph from Etzold).

discuss only two representative and well-documented examples, *Phycomyces* and *Avena*.

12.2.1.1 *PHYCOMYCES*

The coenocytic non-septated hyphae of this saprophytic mold grow horizontally on a substratum; only the sporangia carrying aerial hyphae rise vertically, orienting with negative geotropism. About 24 h after the induction, the tip starts to differentiate into a sporangium. After its maturation the sporangiophore grows with a constant velocity of about 50 μm/min until it reaches a total length of more than 10 cm. The growth is restricted to a zone of 2 to 3 mm immediately under the sporangium. During its growth the sporangiophore can be stimulated phototropically, and bends toward the light source with an angular velocity of about 3°/min, which depends on the differential growth of the two flanks.

This differential growth is induced by an internal light gradient which is caused by a lens effect. The sporangiophore has a higher internal refractive index ($n = 1.36$) than the surrounding air ($n = 1.00$). Thus, parallel light incident from a side is focussed onto the rear side. This lens hypothesis, which we have already discussed for the pseudoplasmodia of the slime mold *Dictyostelium* (cf. Section 12.1.1), has been supported by an inversion experiment. When immersed in a medium with a higher refractive index such as paraffin oil ($n = 1.45$) the normal, convergent ray pattern is altered to a divergent ray pattern. Consequently the fluence rate is higher on the proximal side than on the distal, and the sporangiophore bends away from the light source. The same effect can be induced by placing a glass rod at a short distance in front of the sporangiophore so that the light beam is focussed onto the front wall. In this case the response is disturbed by an avoidance response, since sporangiophores bend away from objects in their vicinity.

Phototropism in *Phycomyces* (Fig. 12.14) is directly linked with a second light-dependent reaction, the light growth response. When the sporangiophore is symmetrically irradiated from two opposite sides with a constant fluence rate, a constant growth rate is observed at fluence rates between 0 and 1 W m^{-2} (provided all other parameters are also kept constant). A sudden increase in fluence rate induces an increase in growth rate after a latency of about 6 min. The sporangiophore adapts to the new fluence rate and then returns to the previous constant growth rate. A sudden decrease in fluence rate induces a transient reduction of growth rate.

When we try to explain phototropism by the phenomenon of the light growth response we are faced with an apparent contradiction. The light growth response is completed within 20 to 30 min, and its amplitude is

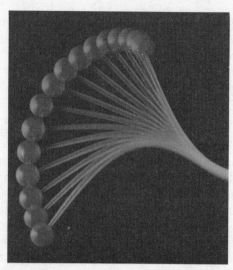

Fig. 12.14. Phototropic bending of the sporangiophore of *Phycomyces blakesleeanus* in the direction of the blue light incident from the left. The multiple exposure was done at 2 min intervals (original photograph courtesy of D. Dennison).

independent of fluence rate. Contrarily, phototropic bending can continue for hours and the bending angle depends on the fluence rate. The maximal angle, however, is about 60° at most fluence rates, due to the antagonistic action of positive phototropism and negative geotropism.

This contradiction can be solved when we consider the rotation of the sporangiophore of about twice an hour. In lateral light a wall element is rotated out of the darker lateral zone into the focussed light band, which induces an increase in the growth rate of this wall element. The photoreceptor is thought to be located in the peripheral cytoplasm or the plasmalemma, which participates in the rotation. While the growth response slows down, the irradiated wall element is rotated out of the focal band and the following, still dark-adapted element is moved into the light beam. As a consequence of this idea we should expect that increased growth should not take place right in the focal band, but when the wall element has been moved a little further due to the latency of the light growth response. In fact the sporangiophore does not bend directly toward the light source but at about 7° to the right of the light source.

When the rotation of the sporangiophore is compensated by placing it on a turning table rotating counterclockwise with the same velocity, phototropic bending is almost inhibited. This supports the hypothesis discussed above, since under this experimental condition the wall elements in the growth zone stay in the same place with respect to light direction and experience no change in fluence rate which could cause a change in growth rate.

The action spectrum of *Phycomyces* sporangiophore phototropism indicates the activity of a typical blue light photoreceptor (Fig. 12.15) similar to the one involved in *Euglena* phototaxis (cf. Section 12.1.1). The spectrum with three maxima in the blue region between 400 and 500 nm (maximum at 450 nm and smaller peaks or shoulders at about 430 nm and 480 nm) could indicate the activity of carotenoids. The additional maximum at 380 nm contradicts this hypothesis, since carotenoids do not absorb in the near UV. An alternative is flavins, which absorb both in the blue and the near UV band. Riboflavin, however, shows only one peak, instead of three peaks, in the blue region.

The discussion concerning the possible photoreceptor has been kept alive by a number of arguments in favor of one or the other alternative (cf. Section 10.3). The flavin hypothesis is supported by the fact that carotenoid free mutants (car⁻) respond phototropically despite their lack of carotenoids. Theoretical considerations, however, show that the amount of carotenoids necessary for a successful photoperception may be too low to be detected analytically or spectrophotometrically.

Recently, the research group of the late Nobel Prize winner, Delbrück, has found a new extremely weak maximum in the action spectrum of the light growth response of *Phycomyces* near 600 nm. The quantum yield of this peak is 10^9 time lower than that of the blue maximum at 455 nm. This

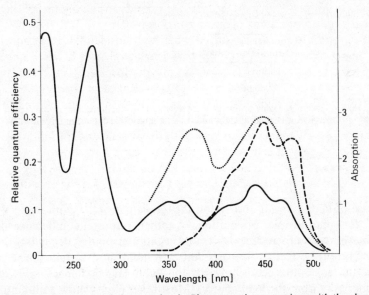

Fig. 12.15. Action spectrum of phototropism in *Phycomyces* in comparison with the absorption spectra (right ordinate) of riboflavin (dotted line) and β-carotene in hexane (dashed line) (from Nultsch and Dennison).

secondary peak could be attributed to the 'forbidden' triplet excitation of riboflavin.

In the spectral range below 300 nm, negative phototropic reactions with a high quantum yield have been observed. The vacuole of the sporangiophore contains gallic acid, which strongly absorbs in this spectral region and prevents any measurable lens effect. Thus the front side is brighter than the rear, and the higher growth rate at the front causes a bending away from the light source.

During the search for photoreceptor mutants a number of other mutants have been isolated which carry lesions in the sensory transduction chain (Fig. 12.16). Comparison of these mutants (called madA, madB, etc.) shows that responses to different external stimuli employ elements of the same transduction chain. One group of the mutants (madA, madB and madC) are 'night-blind'; they need $10-10^6$ times higher fluence rates than the wild type to respond phototropically. The second group is 'stiff', and does not bend as strongly toward the light source.

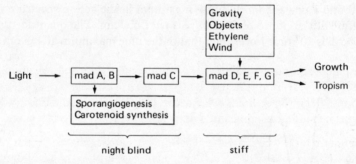

Fig. 12.16. Sequence of the gene products in the sensory transduction chain of some reactions in *Phycomyces*, elicited by light and other stimuli (from Russo, modified).

12.2.1.2 *AVENA*

Because of high internal absorption the *Avena* coleoptile cannot use a lens effect for phototropic orientation. Another important difference from *Phycomyces* is that there are positive and negative responses, depending on the fluence rate (Fig. 12.17). Above a zero threshold, in a certain fluence rate range, the coleoptile bends towards the light. Higher fluence rates cause bending away from the light source. (Most other plants show a minimum in the fluence rate response curve rather than negative responses.) Still higher fluence rates again induce positive curvatures.

To be more precise, the bending angle – at least in the first positive

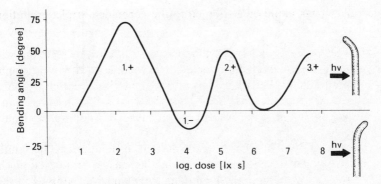

Fig. 12.17. Dependence of the bending angle in phototropism of the *Avena* coleoptile on the dose (after du Buy and Nuernbergh, from Haupt).

curvature – depends not on the fluence rate, but on the product of fluence rate I and the exposure time t (dose). Thus the response quantitatively depends on the number of incident quanta over a wide range (Bunsen–Roscoe law or law of reciprocity).

$$R = c \times I \times t$$

The plant responds with the same bending angle when, e.g., an illuminance of 6.1×10^{-4} lx is irradiated for 10 h ($I \times t = 21.96$ lx s) or 1255 lx for 1.75 ms. Reciprocity holds over a certain – in this example fairly wide – range for both factor time and factor illuminance.

The optimum for the first positive reaction is found at a dose between 100 and 1000 lx s. Higher doses (10^4 lx s) induce negative responses and subsequently the second positive curvature. The dose–response curve of corn shows a similarly complicated behavior without exhibiting a negative response.

The photoperception is restricted to the extreme tip of the coleoptile and the action spectrum indicates the activity of a typical blue light receptor.

The curvature could be initiated by a growth inhibition of the flank facing the light or an enhanced growth on the far side. There is strong indication that the difference in growth rate is controlled by different auxin concentrations (indole-3-acetic acid). This growth hormone is produced in the tip of the coleoptile and transported basally, where it induces growth in the elongated zone. The transported auxin can be collected from a decapitated coleoptile tip into an agar cube and assayed quantitatively by radioactive labelling (^{14}C) or by a biological test (e.g. hypocotyl test). When the auxin is collected separately from the irradiated and dark side of a coleoptile illuminated by a lateral light beam more growth hormone is found in the dark half (Fig. 12.18a). Several hypotheses have been proposed to explain this effect:

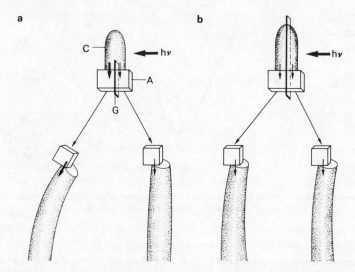

Fig. 12.18. Light-induced auxin transport in the *Avena* coleoptile. (a) Under the influence of light the auxin produced in the coleoptile tip (C) is diverted to the side facing away from the light source so that more auxin diffuses into the left agar block than into the right one (A). The two blocks are separated by a barrier (G). Due to the higher auxin content the left agar block causes a stronger bending when placed on a decapitated coleoptile than the right block. (b) When the lateral transport is blocked by a barrier the same amount of auxin diffuses into both agar blocks (after Nultsch and Haupt, modified).

1. Light could inactivate auxin more on the bright than on the dark side. In fact, auxin is photooxidized in the presence of flavins, but the fluence rates necessary for this photooxidation are much higher than those needed to induce the first positive or negative curvature.

2. The auxin transport could be delayed on the irradiated flank. The photooxidation of auxin produces 3-methylenoxindol, which effectively inhibits auxin transport by impairing the membrane-bound permeases.

3. Light could induce a lateral auxin transport to the dark side. This hypothesis is supported by the fact that no concentration differences are found between the front and rear side when the two flanks are separated by a thin glass barrier.

These explanations have been criticized recently, since only the second positive bending is inhibited by blocking the lateral transport and not the first. Furthermore, phototropic curvatures have been found in which no auxin gradients could be detected. Finally, the bending commences earlier than an effective gradient could be produced by a differential auxin transport. Alternatively, a light-dependent lateral transport of cofactors is being discussed as well as the light effect on auxin already present in the cell. Obviously it is important to study the biochemical mechanism of auxin.

Current research concentrates on binding to membranes, effect on ion pumps, enzyme activity and activation of genes.

12.2.2   Photonastic responses

As described in the introduction to this chapter, phototropic movements depend on the stimulus direction while in photonastic responses the direction of bending is predefined by the morphology of the organ. The movement of stomata, for example, is restricted to an opening and closing of the central pore. The mechanics is based on turgor changes in the guard cells in combination with differential wall thickening, which allows bending only in a predefined direction (Fig. 12.19). During water deficiency the bean-shaped guard cells of *Iris* and related species are less than fully turgescent and therefore close the pore. With increasing turgor pressure the guard cells can only bulge out in the direction of the subsidiary cells, due to the thicker, less elastic front walls and additional cuticular thickenings.

The mechanism is controlled by a complicated circuit in which light is only one external factor in addition to temperature, water potential and the CO_2 concentration. Light causes an opening of the stomata provided all other factors are kept constant. The action spectrum has maxima in the red and blue regions and indicates the activity of the photosynthetic apparatus. This assumption is supported by the fact that guard cells contain functioning

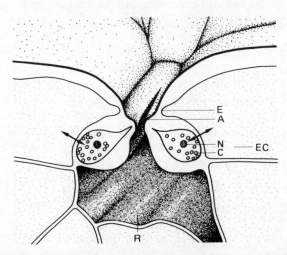

Fig. 12.19. Three-dimensional view of the stomatal apparatus in *Iris*. E = Epidermal joint, A = stomatal aperture, N = nucleus, C = chloroplast, EC = epidermal cell, R = respiratory cavity. The arrows indicate the direction of movement (drawing after microphotographs by Nultsch and Grahle and a sketch by Schuchart).

chloroplasts while the surrounding epidermal cells usually do not. However, blue light has a stronger activity than the absorption spectrum suggests; therefore the existence of an additional blue light receptor has been suggested.

One of the models currently being discussed assumes that the intracellular and intercellular CO_2 concentration decreases in light due to the photosynthetic CO_2 fixation. The resulting CO_2 depletion causes an opening of the stomata.

The following observation supports the hypothesis that light indirectly controls the stomata movement by a change in the actual CO_2 concentration. At a fluence rate of 10 to 25% of full sunlight the CO_2 concentration in the intercellular system of the leaf approaches a minimal value, below which it never falls, even at higher fluence rates. Likewise an increase in the fluence rate up to the critical level causes a proportional opening of the stomata, but even higher fluence rates have no further effect. Furthermore, the stomata stay closed in light when the photosynthetic CO_2 fixation is impaired by exposure to inhibitors of the electron transport chain, e.g. DCMU or cyanazine. A direct light effect on the stoma opening is also being discussed.

An as yet unknown sensor detects the CO_2 concentration and controls an active, energy-consuming transport of osmotically active substances such as the ions K^+, Cl^- and malate between the guard cells and the adjacent cells (Fig. 12.20). In other words, light causes a decrease in CO_2 concentration, which triggers the transport of ions from the adjacent cells into the guard cells.

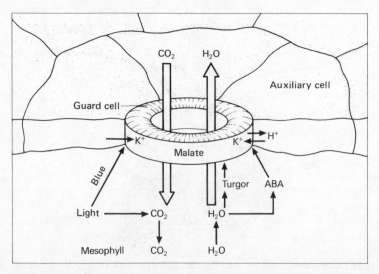

Fig. 12.20. Model of the regulation mechanism in stomates. Blue light, CO_2 concentration and water potential influence the opening of stomates. The mechanics depend on a turgor change induced by ionic currents (from Dittrich).

The potassium influx follows an electrical potential gradient which is produced by an ATP-dependent proton pump (which pumps H^+ out of the guard cells). It is possible that the necessary ATP is produced by cyclic photophosphorylation in the guard cell chloroplasts which contain only photosystem II. Chloride and malate anions compensate the electrical charge of the potassium cations. Malate is produced by carboxylation of phosphoenolpyruvate in a pathway which has been described for C_4 plants (cf. Section 9.6.3.2). PEP carboxylase has been found in guard cells. In addition to the photosynthetic CO_2 fixation in the mesophyll cells, it reduces the CO_2 concentration in the vicinity of the stomata.

The decreasing turgor in the guard cells causes a water uptake from the subsidiary cells. The resulting volume increase of the guard cells causes an opening of the pore in low CO_2 concentration and sufficient water supply. According to another hypothesis the accumulation of photosynthetic intermediate products such as glycolate is responsible for the stoma opening.

Totally independent of the control mechanism discussed above, there is a second mechanism which uses the water potential of the surrounding tissue as an input signal. In this regulation circuit the plant hormone, abscisic acid, acts as messenger and controls the ion transport. This pathway allows the stomata to close during a water stress situation independent of CO_2 concentration. Thus the stoma movement is controlled by a complicated system of feedback mechanisms.

12.3 INTRACELLULAR MOVEMENTS

Light-induced movements of the cytoplasm and intracellular organelles are called photodinesis. Light can elicit or accelerate plasma streaming and induce or control the independent movement of organelles such as the nucleus or plastids. In some cases these two phenomena cannot easily be separated, since some organelles do not move actively but are carried by the plasma streaming.

A typical example of light-induced plasma streaming is found in *Vallisneria*. The cytoplasmic layer in the brick-shaped mesophyll cells rotates, and this can easily be observed since the chloroplasts are passively transported (Fig. 12.21). After an extended dark period the streaming stops. A closer look, however, shows that the inner plasma layers continue to rotate while the outer layers holding the chloroplasts stop. Light activates the outer layers so that the chloroplasts start moving. The action spectrum indicates the activity of the familiar blue light receptor. In addition there is a secondary peak in the red region, which could indicate the participation of chlorophyll. Red light is less effective than blue light. Centrifugation experiments suggest that the chloroplasts are anchored in the cortical plasm (by fibrils?) and that light

GP–J

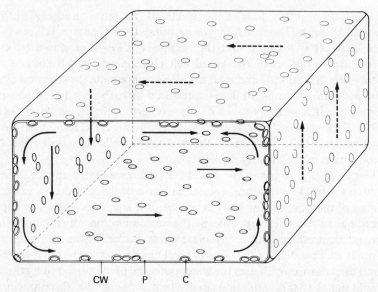

Fig. 12.21. Rotational movement of the plasma layer (P) with the chloroplasts (C) in a mesophyll cell of *Vallisneria*. The front wall is removed and allows the view into the cell. CW = cell wall. The arrows indicate the direction of streaming. The zones moving in opposite directions are separated by a stationary indifference zone.

loosens these connections. Thus, in light, the chloroplasts are transported by the rotating cytoplasm.

The position of plastids of many algae, liverworts, mosses, ferns and higher plants is controlled by the fluence rate and direction of the incident radiation. Most research has concentrated on the filamentous green alga *Mougeotia* and the moss *Funaria hygrometrica*.

The trichomes of *Mougeotia* consists of barrel-shaped cells with one flat chloroplast. In weak light all plastids turn perpendicular to the incident beam (face position). In strong light the edge faces the light source (profile position) (Fig. 12.22). While the strong light orientation in *Mougeotia* is induced by a blue light receptor the weak light orientation is controlled by the phytochrome system (cf. Section 10.2).

The weak light (face) position can be induced by a light pulse of only a few seconds, or even milliseconds. The movement, however, first occurs after about 30 min but can also take place in darkness. The photoreceptor pigments are located in the outer cytoplasmic layers and may be bound to the plasmalemma. Experiments in polarized light suggest that the phytochrome molecules are oriented helically and parallel to the cell surface when in their red-absorbing form. Molecules can absorb photons preferentially when they are oriented more or less perpendicular to the incident light beam. Therefore,

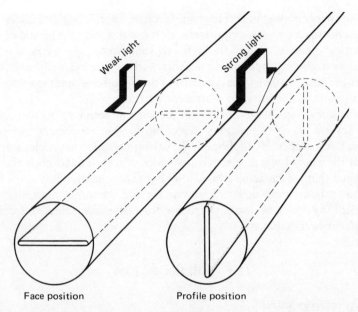

Fig. 12.22. Orientation of the *Mougeotia* chloroplast in weak and strong light (from Haupt, modified).

only those molecules which are adjacent to parts of the cell wall facing the light source and those opposite, are excited. Those at the flanks absorb only a few quanta, since they are oriented more or less parallel to the actinic beam. This symmetry generates a phytochrome gradient. In the absorbing zones the red light-absorbing P_r form is converted into the far red-absorbing P_{fr} form. The plastid orients in this gradient so that its edges point to areas of highest P_r concentration. The induction by blue light could also be mediated by phytochrome (which also absorbs in this spectral region) or an additional blue light photoreceptor.

In *Mougeotia* the action spectra for movement into the face and profile position differ considerably, which is in contrast to *Funaria* and higher plants. The strong light response is elicited by blue light. When a P_{fr} gradient is produced in red light, and the cell is simultaneously hit by blue light, the edges of the chloroplast point to areas of the highest P_{fr} concentration rather than to the lowest, as in the weak light response. Thus, blue light acts as a trigger which switches the response from weak to strong light orientation. Unlike photodinesis in *Vallisneria*, the photosynthetic apparatus is not involved in the photoperception in *Mougeotia*, but only supplies the necessary energy for movement in the form of ATP produced by photophosphorylation (in addition to oxidative phosphorylation).

The active chloroplast movement is brought about by contractile filaments which probably consist of actomyosin. Electron microscopy and interference contrast microscopy show cytoplasmic strands anchoring the plastid edges to the cortical plasma. Cytochalasin B, an inhibitor of the actin filaments, suppresses the weak light movement as well as N-ethylemaleimid, which impairs the function of myosin, sypporting the hypothesis that the actomyosin system generates the force for plastid movements.

The sensory transduction of plastid movements seems to involve calcium ions. Several days of calcium depletion inhibit movement reversibly. Radioactive isotopes (^{45}Ca) showed that calcium is stored in vesicles along the edge of the plastid where it is released during light activation. It could be speculated that the calcium released after light stimulation binds to the fibrils and thus produces movement. Likewise the contraction of muscle fibrils is controlled by calcium ions, which in this case are released from the sarcoplasmic reticulum.

12.4 BIBLIOGRAPHY

Textbooks and review articles

Cerda-Olmedo, E. Genetic determination of the responses of *Phycomyces* to light. In: Senger, H. (ed.) *Blue Light Effects in Biological Systems*. Springer, Berlin, Heidelberg, New York and Tokyo, pp. 220–227 (1984)

Dennison, D. S. Phototropism. In: Haupt, W. and Feinleib, M. E. (eds) *Encyclopedia of Plant Physiology*, Vol. 7: *Physiology of Movements*. Springer, Berlin and Heidelberg, pp. 506–566 (1979)

Diehn, B. Sensory transduction in *Euglena*. In: Colombetti, G., Lenci, F. and Song, P.-S. (eds) *Sensory Perception and Transduction in Aneural Organisms*. Plenum Press, New York and London, pp. 165–178 (1985)

Feinleib, M. E. Behavioral studies of free-swimming photoresponsive organisms. In: Colombetti, G., Lenci, F. and Song, P.-S. (eds) *Sensory Perception and Transduction in Aneural Organisms*. Plenum Press, New York and London, pp. 119–146 (1985)

Firn, R. D. and Digby, J. The establisment of tropic curvatures in plants. *Ann. Rev. Plant Physiol.* **31**, 131–148 (1980)

Foster, K. W. and Smyth, R. D. Light antennas in phototactic algae. *Microbiol. Rev.* **44**, 572–630 (1980)

Häder, D.-P. Photomovement. In Haupt, W. and Feinleib, M. E. (eds) *Encyclopedia of Plant Physiology*, Vol. 7: *Physiology of Movements*. Springer, Berlin and Heidelberg, pp. 268–309 (1979)

Häder, D.-P. Wie orientieren sich Cyanobakterien im Licht? *BIUZ* **14**, 78–83 (1984)

Häder, D.-P. Photomovement. In: Senger, H. (ed.) *Blue Light Effects in Biological Systems*. Springer, Berlin, Heidelberg, New York and Tokyo, pp. 435–443 (1984)

Haupt, W. *Bewegungsphysiologie der Pflanzen*. Thieme, Stuttgart (1977)

Haupt, W. Light mediated movement of chloroplasts. *Ann. Rev. Plant Physiol.* **33**, 205–233 (1982)

Haupt, W. Photoreception and photomovement. *Phil. Trans. R. Soc. Lond. B*, **303**, 467–478 (1983)

Humbeck, K. and Senger, H. The blue light factor in sun and shade plant adaptation. In: Senger, H. (ed.) *Blue Light Effects in Biological Systems*. Springer, Berlin, Heidelberg, New York and Tokyo, pp. 344–351 (1984)

Lenci, F. and Colombetti, G. (eds): *Photoreception and Sensory Transduction in Aneural Organisms*. NATO ASI Series, Vol. 33, Series A: Life Sciences, Plenum Press, New York and London (1980)

Lenci, F., Häder, D.-.P. and Colombetti, G. Photosensory responses in freely motile microorganisms. In: Colombetti, G. and Lenci, F. (eds) *Membranes and Sensory Transduction*. Plenum Press, New York and London, pp. 199–229 (1984)

Lipson, E. D., Galland, P. and Pollock, J. A. Blue light receptors in *Phycomyces* investigated by action spectroscopy, fluorescence lifetime spectroscopy, and two-dimensional gel electrophoresis. In: Senger, H. (ed.) *Blue Light Effects in Biological Systems*. Springer, Berlin, Heidelberg, New York and Tokyo, pp. 228–236 (1984)

Melkonian, M. and Robenek, H. The eyespot apparatus of flagellated green algae: A critical review. In: Round, F. E. and Chapman, D. J. (eds) *Progress in Phycological Research*, Vol. 3. Biopress (1984)

Nultsch, W. Photosensing in cyanobacteria. In: Colombetti, G., Lenci, F. and Song, P.-S. (eds) *Sensory Perception and Transduction in Aneural Organisms*. Plenum Press, New York and London, pp. 147–164 (1985)

Poff, K. L. Mechanisms for the measurement of light direction. In: Colombetti, G., Lenci, F. and Song, P.-S. (eds) *Sensory Perception and Transduction in Aneural Organisms*. Plenum Press, New York and London, pp. 251–263 (1985)

Poff, K. L. and Hong, C. B. Photomovement and photosensory transduction in microorganisms. *Photochem. Photobiol.* **36**, 749–752 (1982)

Raschke, K. Movements of stomata. In: Haupt, W. and Feinleib, M. E. (eds) *Encyclopedia of Plant Physiology*, Vol. 7: *Physiology of Movements*. Springer, Berlin and Heidelberg, pp. 383–441 (1979)

Raven, J. A. Do plant photoreceptors act at the membrane level? *Phil. Trans. R. Soc. B*, **303**, 403–417 (1983)

Seitz, K. and Maurer, B. Influence of ATPase inhibitors on light-dependent movement of chloroplasts in *Vallisneria*. In: Senger, H. (ed.) *Blue Light Effects in Biological Systems*. Springer, Berlin, Heidelberg, New York and Tokyo, pp. 460–462 (1984)

Shropshire, W., Jr. Signal processing in the transduction mechanisms of phototropism. In: Colombetti, G., Lenci, F. and Song, P.-S. (eds) *Sensory Perception and Transduction in Aneural Organisms*. Plenum Press, New York and London, pp. 211–229 (1985)

Smith, H. Plants that track the sun. *Nature* **308**, 774 (1984)

Wagner, G. Actomyosin as a basic mechanism of movement in animals and plants. In: Haupt, W. and Feinleib, M. E. (eds) *Encyclopedia of Plant Physiology*, Vol. 7: *Physiology of Movements*. Springer, Berlin and Heidelberg, pp. 114–126 (1979)

Wagner, G. and Grolig, F. Molecular mechanisms of photoinduced chloroplast movements. In: Colombetti, G., Lenci, F. and Song, P.-S. (eds) *Sensory Perception and Transduction in Aneural Organisms*. Plenum Press, New York and London, pp. 281–298 (1985)

Walczak, T., Gabrys, H. and Haupt, W. Flavin-mediated weak-light chloroplast movement in *Mougeotia*. In: Senger, H. (ed.) *Blue Light Effects in Biological Systems*. Springer, Berlin, Heidelberg, New York and Tokyo, pp. 454–459 (1984)

Walczak, T., Zurzycki, J. and Gabrys, H. Chloroplast displacement response to blue light pulses. In: Senger, H. (ed.) *Blue Light Effects in Biological Systems*. Springer, Berlin, Heidelberg, New York and Tokyo, pp. 444–453 (1984)

Further reading

Armitage, J. P., Ingham, C. and Evans, M. C. W. Role of proton motive force in phototactic and aerotactic responses of *Rhodopseudomonas sphaeroides*. *J. Bact.* **161**, 967–972 (1985)

Berg, H. C. Bovine-like rhodopsin in algae. *Nature* **311**, 702 (1984)

Bilderback, D. E. Phototropism of *Selanginella*: the role of the small dorsal leaves and auxin. *Am. J. Bot.* **71**, 1330–1337 (1984)

Bilderback, D. E. Phototropism of *Selaginella*: the differential response to light. *Am. J. Bot.* **71**, 1323–1329 (1984)

Bittisnich, D. and Williamson, R. E. Control by phytochrome of extension growth and polarotropism in chloronemata of *Funaria*. *Photochem. Photobiol.* **42**, 429–436 (1985)

Blaauw, O. H. and Blaauw-Jansen, G. The phototropic response of *Avena* coleoptiles. *Acta Bot. Neerl.* **19**, 755–763 (1970)

Blatt, M. R. The action spectrum for chloroplast movements and evidence for blue-light-photoreceptor cycling in the alga *Vaucheria*. *Planta* **159**, 267–276 (1983)

Briggs, W. R. and Iino, M. Blue-light-absorbing photoreceptors in plants. *Phil. Trans. R. Soc. B*, **303**, 347–359 (1983)

Cohen, J. D. and Badurski, R. S. Chemistry and physiology of bound auxins. *Ann. Rev. Plant Physiol.* **33**, 403–430 (1982)

Dohrmann, U. In-vitro riboflavin binding and endogenous flavins in *Phycomyces blakesleeanus*. *Planta* **159**, 357–365 (1983)

Eamus, D. and Wilson, J. M. A model for the interaction of low temperature, ABA, IAA, and CO_2 in the control of stomatal behavior. *J. Exp. Bot.* **35**, 91–98 (1984)

Ellis, R. J. Kinetics and fluence-response relationships of phototropism in the dicot *Fagopyrum esculentum Moench*. (Buckwheat). *Plant Cell Physiol.* **25**, 1513–1520 (1984)

Engeln, H., Krause, J., Wachmann, E. and Köhler, W. Negative phototaxis in *Drosophila* associated with a morphological change in the compound eye. *Experientia* **41**, 611–612 (1985)

Farquhar, G. D. and Sharkey, T. D. Stomatal conductance and photosynthesis. *Ann. Rev. Plant Physiol.* **33**, 317–345 (1982)

Foster, K. W., Saranak, J., Patel, N., Zarilli, G., Okabe, M., Kline, T. and Nakanishi, K. A rhodopsin is the functional photoreceptor for phototaxis in the unicellular eukaryote *Chlamydomonas. Nature* **311**, 756–759 (1984)

Gabrys, H., Walczak, T. and Haupt, W. Blue-light-induced chloroplast orientation in *Mougeotia*. Evidence for a separate sensor pigment besides phytochrome. *Planta* **160**, 21–24 (1984)

Galland, P. Action spectra of photogeotropic equilibrium in *Phycomyces* wild type and three behavioral mutants. *Photochem. Photobiol.* **37**, 221–228 (1983)

Galland, P. and Lipson, E. D. Photophysiology of *Phycomyces blakesleeanus*. *Photochem. Photobiol.* **40**, 795–800 (1984)

Galland, P. and Lipson, E. D. Action spectra for phototropic balance in *Phycomyces blakesleeanus*: dependence on reference wavelength and intensity range. *Photochem. Photobiol.* **41**, 323–329 (1985)

Galland, P. and Lipson, E. D. Modified action spectra of photogeotropic equilibrium in *Phycomyces blakesleeanus* mutants with defects in genes madA, madB, madC, and madH. *Photochem. Photobiol.* **41**, 331–335 (1985)

Galland, P. and Russo, V. E. A. Light and dark adaptation in *Phycomyces* phototropism. *J. Gen. Physiol.* **84**, 101–118 (1984)

Ghetti, F., Colombetti, G., Lenci, F., Campani, E., Polacco, E. and Quaglia, M. Fluorescence of *Euglena gracilis* photoreceptor pigment: an in vitro microspectrofluorometric study. *Photochem. Photobiol.* **32**, 29–33 (1985)

Göring, H., Koshuchowa, S., Münnich, H. and Dietrich, M. Stomatal opening and cell enlargement in response to light and phytohormone treatments in primary leaves of red-light-grown seedlings of *Phaseolus vulgaris L. Plant Cell Physiol.* **25**, 638–690 (1984)

Häder, D.-P. Negative phototaxis of *Dictyostelium discoideum* pseudoplasmodia in UV radiation. *Photochem. Photobiol.* **41**, 225–228 (1985)

Häder, D.-P. and Burkart, U. Optical properties of *Dictyostelium discoideum* pseudoplasmodia responsible for phototactic orientation. *Exp. Mycol.*, **7**, 1–8 (1983)

Häder, D.-P. and Wenderoth, K. Role of three basic light reactions in photomovement of desmids. *Planta* **137**, 207–214 (1977)

Harding, R. W. and Melles, S. Genetic analysis of phototropism of *Neurospora crassa* perithecial beaks using white collar and albino mutants. *Plant Physiol.* **72**, 996–1000 (1983)

Haupt, W. Wavelength-dependent action dichroism: A theoretical consideration. *Photochem. Photobiol.* **39**, 107–110 (1984)

Haupt, W., Hupfer, B. and Kraml, M. Blitzlichtinduktion der Chloroplastenbewegung bei *Mougeotia*: Wirkung unterschiedlicher Spektralbereiche und Polarisationsrichtungen. *Z. Pflanzenphysiol.* **96**, 331–342 (1980)

Hertel, R. The mechanism of auxin transport as a model for auxin action. *Z. Pflanzenphysiol.* **112**, 53–67 (1983)

Iino, M. and Briggs, W. R. Growth distribution during first positive phototropic curvature of maize coleoptiles. *Plant Cell Environ.* **7**, 97–104 (1984)

Iino, M. and Schäfer, E. Phototropic response of the stage I *Phycomyces* sporangiophore to a pulse of blue light. *Proc. Natl. Acad. Sci.* **81**, 7103–7107 (1984)

Iino, M., Briggs, W. R. and Schäfer, E. Phytochrome-mediated phototropism in maize seedling shoots. *Planta* **160**, 41–51 (1984)

Jenkins, G. I. and Cove, D. J. Phototropism and polarotropism of primary chloronemata of the moss *Physcomitrella patens*: response of the wild-type. *Planta* **158**, 357–364 (1983)

Kachar, B. Direct visualization of organelle movement along actin filaments dissociated from characean algae. *Science* **227**, 1355–1357 (1985)

Kadota, A., Koyama, M., Wada, M. and Furuya, M. Action spectra for polarotropism and phototropism in protonemata of the fern *Adiantum capillus-veneris*. *Physiol. Plant.* **61**, 327–330 (1984)

Kamiya, R. and Witman, G. B. Submicromolar levels of calcium control the balance of beating between the two flagella in demembranated models of *Chlamydomonas*. *J. Cell Biol.* **98**, 97–107 (1984)

Koga, K., Sato, T. and Ootaki, T. Negative phototropism in the piloboloid mutants of *Phycomyces blakesleeanus Bgff*. *Planta* **162**, 97–103 (1984)

Kraml, M., Enders, M. and Bürkel, N. Kinetics of the dichroic reorientation of phytochrome during photoconversion in *Mougeotia*. *Planta* **161**, 216–222 (1984)

Kumon, K. and Suda, S. Changes in extracellular pH of the motor cells of *Mimosa pudica L.* during movement. *Plant Cell Physiol.* **26**, 375–377 (1985)

Kumon, K. and Tsurumi, S. Ion efflux from pulvinar cells during slow downward movement of the petiole of *Mimosa pudica L.* induced by photostimulation. *J. Plant Physiol.* **115**, 439–443 (1984)

Lipson, E. D., Lopez-Diaz, I. and Pollock, J. A. Mutants of *Phycomyces* with enhanced tropisms. *Exp. Mycol.* **7**, 241–252 (1983)

Macleod, K., Digby, J. and Firn, R. D. Evidence inconsistent with the Blaauw model of phototropism. *J. Exp. Bot.* **36**, 312–319 (1985)

Morel-Laurens, N. M. L. and Feinleib, M. E. Photomovement in an 'eyeless' mutant of *Chlamydomonas*. *Photochem. Photobiol.* **37**, 189–194 (1983)

Nultsch, W. and Schuchart, H. Photomovement of the red alga *Porphyridium cruentum* (Ag.) Naegeli. II. Phototaxis. *Arch. Microbiol.* **125**, 181–188 (1980)

Nultsch, W. and Schuchart, H. A model of the phototactic reaction chain of the cyanobacterium *Anabaena variabilis*. *Arch. Microbiol.* **142**, 180–184 (1985)

Outlaw, W. H., Jr. Current concepts on the role of potassium in stomatal movements. *Physiol. Plant.* **59**, 302–311 (1983)

Paietta, J. and Sargent, M. L. Modification of blue light photoresponses by riboflavin analogs in *Neurospora crassa*. *Plant Physiol.* **72**, 764–766 (1983)

Parks, B. M. and Poff, K. L. Phytochrome conversion as an in situ assay for effective light gradients in etiolated seedlings of *Zea mays*. *Photochem. Photobiol.* **41**, 317–322 (1985)

Parsons, A., Mcleod, K., Firn, R. D. and Digby, J. Light gradients in shoots subjected to unilateral illumination – implications for phototropism. *Plant Cell Envir.* **7**, 325–332 (1984)

Poff, K. L. Perception of a unilateral light stimulus. *Phil. Trans. R. Soc. Lond. B*, **303**, 479–487 (1983)

Pollock, J. A. and Lipson, E. D. A flavoprotein in *Phycomyces blakesleeanus* with short fluorescence lifetime. *Photochem. Photobiol.* **41**, 351–354 (1985)

Rao, I. M. and Anderson, M. E. Light and stomatal metabolism. *Plant Physiol.* **71**, 456–459 (1983)

Rubery, P. H. Auxin receptors. *Ann. Rev. Plant Physiol.* **32**, 569–596 (1981)

Rüffer, U. and Nultsch, W. High-speed cinematographic analysis of the movements of *Chlamydomonas*. *Cell Mot.* **5**, 251–263 (1985)

Schuchart, H. and Nultsch, W. Possible role of singlet molecular oxygen in the control of the phototactic reaction sign of *Anabaena variabilis*. *J. Photochem.* **25**, 317–325 (1984)

Serlin, B. S. and Roux, S. J. Modulation of chloroplast movement in the green alga *Mougeotia* by the Ca^{2+} ionophore A23187 and by calmodulin antagonists. *Proc. Natl. Acad. Sci.* **81**, 6368–6372 (1984)

Song, P.-S., Häder, D.-P. and Poff, K. L. Step-up photophobic response in the ciliate *Stentor coeruleus*. *Arch. Microbiol.* **126**, 181–186 (1980)

Steinitz, B., Ren, Z. and Poff, K. L. Blue and green light-induced phototropism in *Arabidopsis thaliana* and *Lactuca sativa L.* seedlings. *Plant Physiol.* **77**, 248–251 (1985)

Takagi, S. and Nagai, R. Light-controlled cytoplasmic streaming in *Vallisneria* mesophyll cells. *Plant Cell Physiol.* **26**, 941–951 (1985)

Vogelmann, T. C. Site of light perception and motor cells in a sun-tracking lupine (*Lupinus succulentus*). *Physiol. Plant.* **62**, 335–340 (1984)

Vogelmann, T. C. and Björn, L. O. Response to directional light by leaves of a sun-tracking lupine (*Lupinus succulentus*). *Physiol. Plant.* **59**, 533–538 (1983)

Wagner, G. and Rossbacher, R. X-ray microanalysis and chlorotetracycline staining of calcium vesicles in the green alga *Mougeotia*. *Planta* **149**, 298–305 (1980)

Wagner, G., Valentin, P., Dieter, P. and Marme, D. Identification of calmodulin in the green alga *Mougeotia* and its possible function in chloroplast reorientational movement. *Planta* **162**, 62–67 (1984)

Walne, P. L., Lenci, F., Mikolajczyk, E. and Colombetti, G. Effect of pronase treatment on step-down and step-up photophobic responses in *Euglena gracilis*. *Cell Biol. Int. Rep.* **8**, 1017–1027 (1984)

Willmer, C. M., Rutter, J. C. and Meidner, H. Potassium involvement in stomatal movements of *Paphiopedilum*. *J. Exp. Bot.* **34**, 507–513 (1983)

13 Molecular basis of vision

This chapter is not intended to give a detailed description of the morphology and anatomy of the different types of eyes found in animals. Neither does this short review cover the optical properties. It rather describes the molecular photochemical and electrophysiological events which cause electrical potential changes in invertebrates and vertebrates, which eventually are relayed as signals along the neurons to the brain.

During cellular sensory transduction we find an extremely large signal amplification. A single photon can induce an electric response in a dark-adapted photoreceptor cell. When we compare the absorbed energy with the electric response an amplification by a factor of several millions is apparent.

13.1 STRUCTURE OF THE EYE AND THE PHOTORECEPTOR CELLS

The eyes of vertebrates and some invertebrates (for instance polychetes and cephalopodes) have lenses which focus the rays onto the retina (Fig. 13.1). Since the retina in the eye of vertebrates (in contrast to the analogously

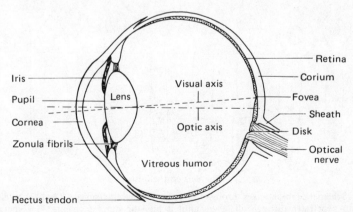

Fig. 13.1. Longitudinal section through a mammalian lens eye (from Hadorn and Wehner and Rodieck, modified).

developed eyes of cephalopods) is formed by an evagination of the forebrain, the photoreceptor cells (rods and cones) are oriented away from the impinging rays (inverse orientation). Therefore the rays first have to pass the covering layers of neurons, bipolar and horizontal cells as well as the inner segments of the rods and cones before they strike the light-sensitive outer segments (Fig. 13.2).

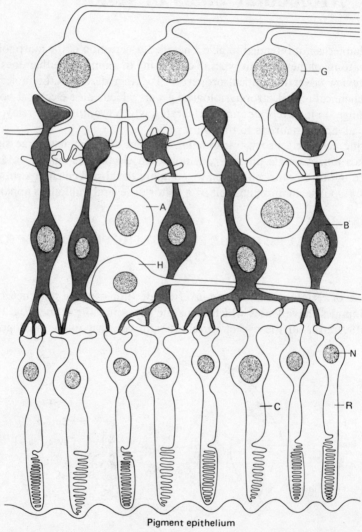

Pigment epithelium

Fig. 13.2. Schematic cross-section through the retina of a mammalian eye. The light enters from above and penetrates through all layers until it hits the outer segments. G=Neural cell, A=amacrine cell, B=bipolar cell, H=horizontal cell, C=cone, R=rod, N=nucleus (from Rodieck and Hadorn and Wehner, modified).

The rods of vertebrates (Fig. 13.3a) are specialized for photoreception in weak light conditions (= skotopic vision). They have an inner segment with a basal synapsis connected to the following bipolar cells and on the other end it is connected with the outer segments by a ciliary bridge. The inner segment contains the nucleus and, among other organelles, many mitochondria which produce the energy needed by the cell. The outer segment is filled with densely packed stacks of closed disks which carry the photoreceptor molecules,

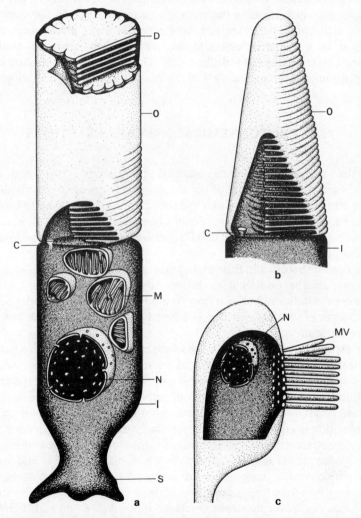

Fig. 13.3. Three-dimensional drawings of (a) a rod and (b) a cone of the frog eye and (c) schematic drawing of an invertebrate visual cell. D = disk, O = outer segment, I = inner segment, C = ciliary bridge, M = mitochondria, N = nucleus, S = synapsis, MV = microvilli (from Rodieck and Stieve, modified).

rhodopsin. These flattened membranous vesicles are formed at the basis by invaginations of the cytoplasmic membrane and are later separated.

The cones of vertebrates (Fig. 13.3b), which are specialized for color vision at sufficient irradiation (photopic vision), differ from the rods, notably in the outer segments. The pigment-bearing membrane does not form closed vesicles but rather invaginations of the cytoplasmic membrane. Both rods and cones also contain rhodopsin in the cytoplasmic membrane.

The outer segments of invertebrate photoreceptor cells carry a large number of microvilli which increase the surface area (Fig. 13.3c). In the compound eyes of arthropods there are normally seven or eight photoreceptor cells arranged in a cylindric ommatidium, with linearly arranged microvilli (rhabdomers) pointing towards the center. The pigment concentration can be very high: in *Astacus* almost 50% of the membrane protein is rhodopsin.

13.2 PHOTOCHEMICAL PRIMARY REACTIONS

More than 100 years ago Kühne isolated the photoreceptor pigment which he called 'Sehpurpur'. In the 1950s the chromophoric group of this rhodopsin, retinal, was identified as 11-*cis* aldehyde of vitamin A, by Wald and Morton. It is linked to a protein (opsin) by a Schiff's base. Several opsins with molecular weights between 35,000 and 50,000 Dalton are found in different animals and different photoreceptor cells.

The rhodopsins differ in their absorption maxima (Fig. 13.4) so that eyes of different species have different spectral sensitivities. True color vision, which in addition to vertebrates has also been found in fishes and insects, depends on the existence of several (often three) different types of cones which absorb in different spectral regions.

The rhodopsin molecule spans the whole membrane and the oligosaccharide side chains of the opsin are oriented inwards towards the vesicle, while the phosphorylation site points outwards (see Section 13.3.2). Absorption of a quantum initiates a whole cycle of processes which includes several intermediates. Dark-adapted rhodopsin from bovine photoreceptor cells has an absorption maximum of 498 nm and the retinal is an 11-*cis* isomer. After light absorption it forms bathorhodopsin (prelumirhodopsin) within 1 ps, which absorbs at 543 nm. This intermediate is stable at − 140 °C, and *in vivo* it reacts within 1 ns to lumirhodopsin. Either during this reaction, or more probably, during the formation of bathorhodopsin, it isomerizes to the all-*trans* form. The following steps to metarhodopsin I (478 nm) and II (380 nm) are connected with a drastic conformational change of the protein which could play an important role during the following steps in the reaction chain (see Section 13.3). Before the receptor potential occurs, a measurable potential

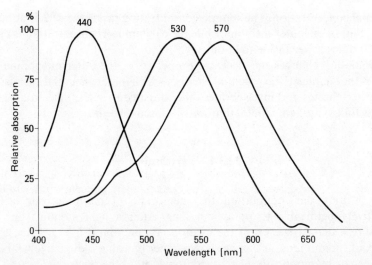

Fig. 13.4. Microspectrophotometrically measured action spectra of isolated human cone outer segments. The rhodopsins (absorbing blue, green and yellow) are not different in their chromophoric group but in their protein component (opsin) (after MacNichol, from Czihak, modified).

change (early receptor potential, ERP) modifies the properties of the membrane in such a way that it causes changes in the ionic currents into the cells. In addition, the molecule binds a proton during the transition from metarhodopsin I to II, as Bennet has shown with pH indicators.

As a last step, in vertebrates the all-*trans* retinal is split hydrolytically from the protein opsin, which can be regenerated to the initial form within a few minutes by reacting with a new 11-*cis* retinal. The unbound retinal absorbs at about 387 nm and therefore light-bleached rhodopsin has a pale yellow appearance. As early as 1878 Kühne described the photobleaching of the 'Sehpurpur'. Before retinal is cleaved from the protein, rhodopsin can be reversed from an intermediate to its initial active form by absorption of a quantum.

13.3 ELECTROCHEMICAL SUBSEQUENT REACTIONS

Like any living cell, each photoreceptor cell has an (internally negative) electrical potential compared to the outside, which can be measured using microelectrodes. This potential is basically a diffusion potential, and is due to differences in the ion concentration between the interior and exterior of the cell. Photoreceptor cells accumulate potassium ions to about 20 times the external concentration. The sodium concentration inside the cells is about 10 times that of the outer medium, while the calcium ion concentration is about 10,000 times lower inside than outside.

Irradiation changes the electrical properties of the outer membrane and causes the characteristic potential changes which initiate neural transmission. Since vertebrates and invertebrates differ drastically in their electrochemical events following light stimulation we will discuss them separately.

13.3.1   Vertebrates

After a long dark-adaptation the membrane of photoreceptor cells is selectively permeable to potassium ions. After a light stimulus one can measure a depolarization of the membrane potential (Fig. 13.5) due to an influx of sodium ions, as can be shown either by using radioactively labelled isotopes or by monitoring the ion composition of the outer medium.

Potential changes are the result of ionic currents; therefore it is often more accurate to measure the actual currents while the potential is kept constant experimentally by a compensating current introduced through a second electrode (voltage clamp).

Absorption of a single photon can increase the membrane conductivity by about 50 nS (1 Siemens = 1/Ohm) which results in a transient potential change of up to 40 mV, called a 'bump'. The amplitude of the bump, and the latency with which it follows the test signal, can vary considerably. Bumps even occur in the absence of light stimuli, which is probably due to thermally induced reactions of rhodopsin.

At higher fluence rates the bumps, caused by single photons, add up to a

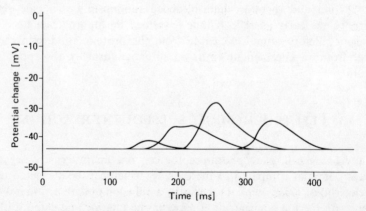

Fig. 13.5. Light-induced potential changes (bumps) in *Limulus* after a 50 ms white light pulse (about 1.26×10^{10} quanta cm^{-2} s^{-1}) (from Stieve).

continuous receptor potential. Both the size of the receptor potential and the membrane conductivity (measured under voltage clamp) depends on the intensity of light stimulus.

Bumps are also influenced by the state of adaptation within the cell. Light adaptation can be caused by screening pigment movement, by neural adaptation of the subsequent synapses and changes in the sensitivity of the photoreceptor cell by several orders of magnitude. This influences both the amplitude and the response time.

One mechanism of adaptation depends on the bleaching of rhodopsin. When at high fluences rates a large fraction of the rhodopsin molecules is bleached the absorption probability for further quanta decreases proportionally. However, even in very strong continuous light only part of the rhodopsin is bleached, since absorption of a second quantum may lead back to the initial state.

The control of the amplification factor within the subsequent electrochemical or biochemical reactions (see below) may play a more important role. Since dark-adapted photoreceptor cells may even respond to the absorption of only one quantum, the cellular transduction has to involve very efficient amplification mechanisms which amplify the signal energy of the absorbed photon about a million-fold. As in the case of microorganisms responding to light, the amplification is effected by gated ion currents through selective membrane channels along previously established ion gradients. Such gradients are formed by energy-consuming ion pumps. These are membrane-bound enzymes which, for example, transport sodium ions in one direction and potassium ions in the opposite, under ATP consumption. They can be impaired by specific blockers such as ouabain or by uncouplers which interrupt the ATP supply of the cell. Arthropod photoreceptor cells lose their light sensitivity when the Na/K pump is blocked. At least one additional pump, which transports calcium ions, has been demonstrated indirectly by removing calcium from the cytoplasm.

The second important membrane component in addition to ion pumps are ion channels, which are either continuously open and allow a constant though selective flux of ions (potassium in invertebrates) or open and close under the control of light or potential changes.

It is feasible that the rhodopsin molecule itself is a selective gate for specific ions and opens by a conformational change. As an alternative it may control adjacent ion channels, which is more probable since absorption of one quantum opens several channels (in the dark-adapted state 1000 to 10,000) which causes a massive ion current. Whichever explanation may turn out to be correct, after a light stimulus we find a gated sodium current into the cell which depolarizes the resting potential (Fig. 13.6).

The depolarization of the cell caused by the light-activated rhodopsin controls voltage-dependent ion channels. First calcium channels are opened

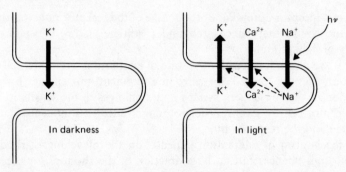

Fig. 13.6. Ion transport through the microvilli membranes of invertebrates in darkness and in light (from Stieve, modified).

which selectively allow a depolarization and thus effect an amplification. Shortly afterwards we find a compensating potassium efflux which repolarizes the membrane, thus terminating the bump. At higher fluence rates the reaction is saturated even before all light-activated ion channels are opened.

After discussing the ion transport phenomena we return to the problem of adaptation. Since bleaching of rhodopsin cannot be the only explanation for adaptation (see above) researchers looked for a mechanism which controls the conductivity of the membrane. If the number of light- and/or voltage-controlled ion channels is decreased this should reduce the light-induced current and thus reduce the amplitude of the receptor potential for the same stimulus. In fact many results indicate a control of membrane permeability by calcium ions, which effectively changes the amplification. Bumps can be suppressed by a large increase in calcium concentration in the outer medium. When calcium ions are almost completely removed from the medium the membrane conductivity increases to a saturation level; at that point no more receptor potentials are possible.

When the calcium concentration is increased by the first light stimulus more ion channels are blocked and the amplification decreases for the following stimuli. During irradiation the intracellular calcium concentration increases; consequently the number of channels opened by light decreases. By this mechanism the sensitivity of the cell can be decreased by a factor of 10,000. In fact, after irradiation an increase of intracellular calcium concentration can be detected with the arsenazo III method (measurement of fluorescent changes). When the intracellular calcium concentration is reduced by injection of EGTA (ethylene glycol bis-aminoethyl ether), a calcium ion chelator, the ability of the cell to adapt is reduced. According to Stieve's hypothesis calcium and sodium ions compete for binding sites on the outside of light-controlled channels; sodium ions open the channels and calcium ions close them. In the unstimulated state (darkness) it is predominantly calcium ions that occupy the binding sites, while in light sodium ions bind and open the channels.

13.3.2 Invertebrates

In vertebrates receptor potentials are generated by a totally different mechanism than in invertebrates. While in invertebrates the absorption of a photon causes a depolarizing bump, light absorption reduces the conductivity in vertebrates' photoreceptor cells, causing hyperpolarization.

In darkness a current (defined as positive charges) flows from the outer segment to the inner segment (Fig. 13.7) where it leaves the cell and returns outside to the outer segment. This current is driven by a potential difference between inner and outer segment. It is interesting to note that different cations are involved in the current: the influx into the outer segment is mainly sodium

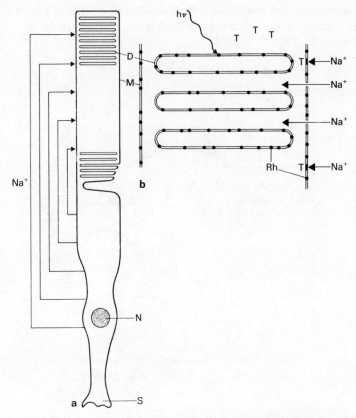

Fig. 13.7. Ionic fluxes in vertebrate rods. (a) In the unstimulated state an external ion current flows from the inner segment to the outer segment. The ion channels are closed in light by a transmitter (Ca^{2+}?) released after absorption of a photon by rhodopsin. (b) Enlarged detail: D = disks, M = membrane, N = nucleus, T = transmitter, Rh = rhodopsin, S = synapsis (from Stieve, modified).

ions, while potassium ions leave the inner segment. In vertebrate photoreceptor cells these ionic currents are also based on concentration gradients which
are generated by energy-consuming ion pumps. Since, at least in rods, the
light-perceiving membrane vesicles are not in contact with the cytoplasmic
membrane the absorption of a quantum needs to be relayed by a transmitter to
modify the electrical properties of the cytoplasmic membrane. According to
the hypothesis of Hagins and his coworkers, as well as Hanava and Matsuura,
this transmitter could consist of calcium ions. The liberated calcium ions are
thought to close the ion channels for the sodium influx which, in contrast to
invertebrate photoreceptor cells, causes a hyperpolarization. Calculated from
the amplification in a visual cell each photon activates more than 1000
transmitters which block the ion channels. However, such a high calcium
concentration could not be demonstrated experimentally within the cells.

An alternative explanation has been developed by Kühn, Liebmann and
Stryer: it involves an enzyme cascade (Fig. 13.8). Absorption of a photon

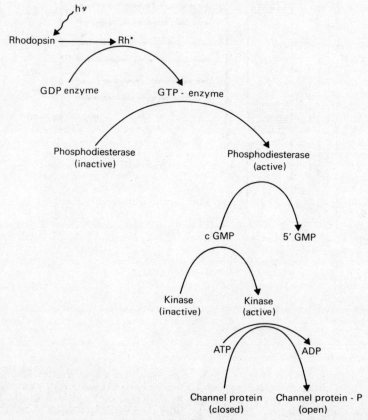

Fig. 13.8. Enzyme cascade amplifying the signal in a vertebrate cell. For details see text (from
Kühn, Liebmann and Stryer).

results in the hydrolysis of a large number of cGMP molecules. The excited rhodopsin (metarhodopsin II?) catalyzes the exchange of GTP for GDP bound to a subunit of transducin, a multisubunit peripheral membrane protein, which is an intermediate in the activation of cGMP phosphodiesterase. Since at low light levels up to 500 enzyme molecules can bind one after the other to the activated rhodopsin molecule, until the latter is deactivated by releasing the retinal, this first step causes an amplification of up to 500-fold.

The GTP enzyme complex activates a cGMP phosphodiesterase which in turn hydrolyzes more than 1000 molecules of cGMP into 5'GMP, giving an overall gain of more than 500,000. The cGMP is supposed to activate a kinase which phosphorylates a membrane protein and opens a sodium channel. The light-induced decrease in cGMP (deactivated by the phosphodiesterase) would thus reduce the activity of the kinase: consequently less membrane proteins are phosphorylated and more channels are closed.

13.4 BIBLIOGRAPHY

Textbooks and review articles

Applebury, M. L. The primary processes of vision: a view from the experimental side. *Photochem. Photobiol.* **32**, 425–431 (1980)

Cervetto, L., McNaughton, P. A., Rispoli, G. and Torre, V. A possible role for calcium and cGMP in rod photoresponse. In: Fein, A. and Levine, J. S. (eds) *The Visual System*. Alan R. Liss, New York, pp. 11–26 (1985)

Chabre, M. From the photon to the neuronal signal. *Europhys. News* **16**, 1–4 (1985)

Chabre, M. Enzymatic amplification mechanism of visual transduction signal in retinal rods. In: Colombetti, G., Lenci, F. and Song, P.-S. (eds) *Sensory Perception and Transduction in Aneural Organisms*. Plenum Press, New York and London, pp. 309–320 (1985)

Chabre, M. Trigger and amplification mechanisms in visual phototransduction. *Ann. Rev. Biophys. Chem.* **14**, 331–360 (1985)

Horridge, G. A. Comparative physiology and evolution of vision in invertebrates. In: Autrum, H. (ed.) *Handbook of Sensory Physiology*, VII/6. Springer, Berlin, Heidelberg and New York (1979)

Kaupp, U. B. The role of calcium in visual transduction. In: Bolis, C. L., Helmreich, E. J. M. and Passow, H. (eds) *Information and Energy Transduction in Biological Membranes*. Alan R. Liss, New York, pp. 325–339 (1984)

Kawamura, S. Involvement of ATP in activation and inactivation sequence of phosphodiesterase in frog rod outer segments. *Biochim. Biophys. Acta* **732**, 276 (1983)

Kühn, H. Early steps in the light-triggered activation of the cyclic GMP enzymatic pathway in rod photoreceptors. In: Bolis, C. L., Helmreich, E. J. M. and Passow, H. (eds) *Information and Energy Transduction in Biological Membranes*. Alan R. Liss, New York, pp. 303–311 (1984)

Kühn, H. and Chabre, M. Light-dependent interactions between rhodopsin and photoreceptor enzymes. *Biophys. Struct. Mech.* **9**, 231–234 (1983)

Miller, W. H. (ed.) *Current Topics in Membranes and Transport*, Vol. 15: *Molecular Mechanisms of Photoreceptor Transduction*. Academic Press, New York (1981)

Miller, W. H., Shimoda, Y. and Hurley, J. B. Cyclic GMP hydrolysis is necessary for rod phototransduction. In: Fein, A. and Levine, J. S. (eds) *The Visual System*. Alan R. Liss, New York, pp. 1–10 (1985)

Packer, L. Biomembranes: visual pigments and purple membranes. In: *Methods in Enzymology*, Vols 81 and 88. Academic Press, New York (1982)

Pappin, D. J. C., Eliopoulos, E., Brett, M. and Findlay, J. B. C. A structural model for bovine rhodopsin. *Int. J. Biol. Macromol.* **6**, 73–76 (1984)

Rodieck, R. W. *The Vertebrate Retina*. W. H. Freeman, San Francisco (1973)

Shichi, H. *Biochemistry of Vision*. Academic Press, Orlando (1983)

Stavenga, D. G. and de Grip, W. J. Progress in phototransduction. *Biophys. Struct. Mech.* **9**, 225–230 (1983)

Stieve, H. Roles of calcium in visual transduction in invertebrates. In: Laverack, M. S. and Cosens, D. J. (eds) *Sense Organs*, Kap. 10. Blackie, Glasgow and London, pp. 163–185 (1981)

Stieve, H. Photorezeption und ihre molekularen Grundlagen. In: Hoppe, W., Lohmann, W., Markl, H. and Ziegler, H. (eds) *Biophysik*, 2nd edn. Springer, Berlin, Heidelberg and New York (1982)

Stieve, H. On the transduction mechanism in the photoreceptor cell of an invertebrate, studied by single photon responses. In: Bolis, C. L., Helmreich, E. J. M. and Passow, H. (eds) *Information and Energy Transduction in Biological Membranes*. Alan R. Liss, New York, pp. 313–324 (1984)

Stryer, L. Light-activated retinal proteins. *Nature* **312**, 498–499 (1984)

Stryer, L. Molecular design of an amplification cascade in vision. *Biopol.* **24**, 29–47 (1985)

Stryer, L., Hurley, J. B. and Fung, K.-K. Transduction and the cyclic phosphodiesterase of retinal rod outer segments. *Meth. Enzymol.* **96**, 617–627 (1983)

Yamazaki, A., Halliday, K. R., George, J. S., Nagao, S., Kuo, C.-H., Ailsworth, K. S. and Bitensky, M. W. Homology between lightactived photoreceptor phosphodiesterase and hormone-activated adenylate cyclase system. In: Cooper, D. M. F. and Seamon, K. B. (eds) *Advances in Cyclic Nucleotide and Protein Phosphorylation Research*, Vol. 19. Raven Press, New York, pp. 113–124 (1985)

Further reading

Alkon, D. L., Farley, J., Sakakibara, M. and Hay, B. Voltage-dependent calcium and calcium-activated potassium currents of a molluscan photoreceptor. *Biophys. J.* **46**, 605–614 (1984)

Andrews, L. D. and Cohen, A. I. Freeze-fracture studies of photoreceptor membranes: new observations bearing upon the distribution of cholesterol. *J. Cell Biol.* **97**, 749–755 (1983)

Azuma, K. and Azuma, M. Absorbance and circular dichroism spectra of 7-cis photoproduct formed by irradiating frog rhodopsin. *Photochem. Photobiol.* **41**, 165–169 (1985)

Azuma, M. and Azuma, K. Chromophore of a long-lived photoproduct formed with metarhodopsin III in the isolated frog retina. *Photochem. Photobiol.* **40**, 495–499 (1984)

Bachhuber, K. and Frösch, D. Electron microscopy of melamine-embedded frog retina: evidence for the overall crystalline organization of photoreceptor outer segments. *J. Microsc.* **133**, 103–109 (1984)

Bacigalupo, J. and Lisman, J. E. Light-activated channels in *Limulus* ventral photoreceptors. *Biophys. J.* **45**, 3–5 (1984)

Bayramashvili, D. I., Drachev, A. L., Drachev, L. A., Kaulen, A. D., Kudelin, A. B., Martynov, V. I. and Skulachev, V. P. Proteinase-treated photoreceptor discs. Photoelectric activity of the partially-digested rhodopsin and membrane orientation. *Europ. J. Biochem.* **142**, 583–590 (1984)

Borys, T. J., Uhl, R. and Abrahamson, E. W. Cyclic GMP stimulation of a light-activated ATPase in rod outer segments. *Nature* **304**, 733–735 (1983)

Brown, H. M. The role of H^+ and Ca^{2+} in *Balanus* photoreceptor function. In: *The Physiology of Excitable Cells*. A. R. Liss, New York, pp. 327–341 (1983)

Brown, J. E. and Rubin, L. J. A direct demonstration that inositol-triphosphate induces an increase in intracellular calcium in *Limulus* photoreceptors. *Biochem. Biophys. Res. Commun.* **125**, 1137–1142 (1984)

Catt, M., Ernst, W., Kemp, C. M. and O'Bryan, P. M. Rhodopsin bleaching and rod adaptation. *Biochem. Soc. Trans.* **11**, 676–678 (1983)

Chen, D.-M., Collins, J. S. and Goldsmith, T. H. The ultraviolet receptor of bird retinas. *Science* **225**, 337–340 (1984)

Clack, J. W., Oakley, II, B. and Stein, P. J. Injection of GTP-binding protein or cyclic GMP phosphodiesterase hyperpolarizes retinal rods. *Nature* **305**, 50–52 (1983)

Cobbs, W. H. and Pugh, E. N., Jr. Cyclic GMP can increase rod outer-segment light-sensitive current 10-fold without delay of excitation. *Nature* **313**, 585–587 (1985)

Doukas, A. G., Junnarkar, M. R., Alfano, R. R., Callender, R. H. and Balogh-Nair, V. The primary event in vision investigated by time-resolved fluorescence spectroscopy. *Biophys. J.* **47**, 795–798 (1985)

Drikos, G., Morys, P. and Rüppel, H. Polarized absorption spectra of monocrystalline all-trans and 11-cis, 12-s-cis retinal at 4.2 K. *Photochem. Photobiol.* **40**, 133–135 (1984)

Fesenko, E. E., Kolesnikov, S. S. and Lyubarsky, A. L. Induction by cyclic GMP of cationic conductance in plasma membrane of retinal rod outer segment. *Nature* **313**, 310–313 (1985)

Foster, R. G., Follett, B. K. and Lythgoe, J. N. Rhodopsin-like sensitivity of extra-retinal photoreceptors mediating the photoperiodic response in quail. *Nature* **313**, 50–52 (1985)

Hanke, W. and Kaupp, U. B. Incorporation of ion channels from bovine rod outer segments into planar lipid bilayers. *Biophys. J.* **46**, 587–595 (1985)

Hargarve, P. A., McDowell, J. H., Curtis, D. R., Wang, J. K., Juszczak, E., Fong, S.-L., Rao, J. K. M. and Argos, P. The structure of bovine rhodopsin. *Biophys. Struct. Mech.* **9**, 235–244 (1983)

Harosi, F. I. and Hashimoto, Y. Ultraviolet visual pigment in a vertebrate: a tetrachromatic cone system in the dace. *Science* **222**, 1021–1023 (1983)

Hofmann, K. P., Emeis, D. and Schnetkamp, P. P. M. Interplay between hydroxylamine, metarhodopsin II and GTP-binding protein in bovine photoreceptor membranes. *Biochim. Biophys. Acta* **725**, 60–70 (1983)

Hurwitz, R. L., Bunt-Milam, A. H., Chang, M. L. and Beavo, J. A. cGMP phosphodiesterase in rod and cone outer segments of the retina. *J. Biol. Chem.* **260**, 568–573 (1985)

Ivens, I. and Stieve, H. Influence of the membrane potential on the intracellular light induced Ca^{2+}-concentration change of the *Limulus* ventral photoreceptor monitored by Arsenazo III under voltage clamp conditions. *Z. Naturf.* **39c**, 986–992 (1984)

Kirschfeld, K., Feiler, R., Hardie, R., Vogt, K. and Franceschini, N. The sensitizing pigment in fly photoreceptors. *Biophys. Struct. Mech.* **10**, 81–92 (1983)

Knowles, A. Rhodopsin bleaching intermediates and enzyme activation in the rod outer segment. *Biochem. Soc. Trans.* **11**, 672–674 (1983)

Korenbrot, J. J. Signal mechanisms of phototransduction in retinal rod. *CRC Crit. Rev. Biochem.* **17**, 223–256 (1984)

Kruizinga, B., Kamman, R. L. and Stavenga, D. G. Laser induced visual pigment coversions in fly photoreceptors measured *in vivo*. *Biophys. Struct. Mech.* **9**, 299–307 (1983)

Kühn, H., Hall, S. W. and Wilden, U. Light-induced binding of 48-kDa protein to photoreceptor membranes is highly enhanced by phosphorylation of rhodopsin. *FEBS Lett.* **176**, 473–478 (1984)

Matthews, H. R., Torre, V. and Lamb, T. D. Effects on the photoresponse of calcium buffers and cyclic GMP incorporated into the cytoplasm of retinal rods. *Nature* **313**, 582–585 (1985)

Miki, N., Kuo, C.-H., Hayashi, Y. and Akiyama, M. Functional role of calcium in photoreceptor cells. *Photochem. Photobiol.* **32**, 503–508 (1980)

Nagy, K. and Stieve, H. Changes in intracellular calcium ion concentration, in the course of dark adaptation measured by Arsenazo III in the *Limulus* photoreceptor. *Biophys. Struct. Mech.* **9**, 207–223 (1983)

Pellicone, C., Nulland, G., Leininger, D. and Virmaux, N. Topologie de la rhodopsine bovine dans les membranes discales des photorecepteurs. *C.R. Acad. Sci. Paris.* **296**, 7–10 (1983)

Pynset, P. B. and Duncan, G. Reconstruction of photoreceptor membrane potentials from simultaneous intracellular and extracellular recordings. *Nature* **269**, 257–259 (1977)

Saibil, H. R. A light-stimulated increase of cyclic GMP in squid photoreceptors. *FEBS Lett.* **168**, 213–216 (1984)

Schwemer, J. and Henning, U. Morphological correlates of visual pigment turnover in photoreceptors of the fly, *Calliphora erythrocephala*. *Cell Tiss. Res.* **236**, 293–303 (1984)

Shinozawa, T. and Bitensky, M. W. Co-operation of peripheral and integral membrane proteins in the light dependent activation of rod GTPase and phosphodiesterase. *Photochem. Photobiol.* **32**, 497–502 (1980)

Stieve, H. and Bruns, M. Bump latency distribution and bump adaptation of *Limulus* ventral nerve photoreceptor in varied extracellular calcium concentrations. *Biophys. Struct. Mech.* **9**, 329–339 (1983)

Stieve, H., Bruns, M. and Gaube, H. The sensitivity shift due to light adaptation depending on the extracellular calcium ion concentration in *Limulus* ventral nerve photoreceptor. *Z. Naturf.* **39c**, 662–679 (1984)

Stockbridge, N. and Ross, W. N. Localized Ca^{2+} and calcium-activated potassium conductances in terminals of a barnacle photoreceptor. *Nature* **309**, 266–268 (1984)

Tokunaga, F., Sasaki, G. and Yoshizawa, T. Orientation of retinylidene chromphore of hypsorhodopsin in frog retina. *Photochem. Photobiol.* **32**, 447–453 (1980)

Tsukahara, Y. Effect of intracellular injection of EGTA and tetraethylammonium chloride on the receptor potential of locust photoreceptors. *Photochem. Photobiol.* **32**, 509–514 (1980)

Vanderberg, C. A. and Montal, M. Light-regulated biochemical events in invertebrate photoreceptors. 1. Light-activated guanosinetriphosphatase, guanine nucleotide binding, and cholera toxin catalyzed labeling of squid photoreceptor membranes. *Biochem.* **23**, 2339–2347 (1984)

Vanderberg, C. A. and Montal, M. Light-regulated biochemical events in invertebrate photoreceptors. 2. Light-regulated phosphorylation of rhodopsin and phosphoinositides in squid photoreceptor membranes. *Biochem.* **23**, 2347–2352 (1984)

Verger-Bocquet, M. Etude infrastructurale des organes photorecepteurs chez les larves de deux Syllidiens (Annelides, Polychetes). *J. Ultrastruc. Res.* **84**, 67–72 (1983)

Vogt, K. Is the fly visual pigment a rhodopsin? *Z. Naturforsch.* **38**, 329–333 (1983)

Waloga, G. and Anderson, R. E. Effects of inositol-1,4,5-trisphosphate injections into salamander rods. *Biochem. Biophys. Res. Commun.* **126**, 59–62 (1985)

Westphal, C., Bachhuber, K. and Frösch, D. Conventional transmission electron microscopy (CTEM) of unstained, melamine-embedded frog photoreceptor membranes. *J. Microsc.* **133**, 111–116 (1984)

Yamamoto, K. and Shichi, H. Rhodopsin phosphorylation occurs at metarhodopsin II level. *Biophys. Struct. Mech.* **9**, 259–267 (1983)

Yau, K.-W. and Nakatani, K. Light-induced reduction of cytoplasmic free calcium in retinal rod outer segment. *Nature* **313**, 579–582 (1985)

14 Biological effects of ultraviolet radiation

The damaging effect of ultraviolet radiation is very obvious from sunburn. It is UV-B radiation, which begins at 285 nm in the natural global radiation, that causes the skin damage (erythema). UV-C radiation, emitted by artificial UV sources (see Section 2.2.2.1), is also biologically very effective: it kills microorganisms and induces mutations and skin cancer. In these cases UV causes genetic damage with nucleic acids as the main targets. In addition, structural and enzyme proteins absorb in the UV range. In plants some growth regulators and pigments absorb in the UV range (Fig. 14.1). However, not all UV-absorbing biomolecules need to be the primary effector sites. Energy transfer to other molecules makes the distinction between primary and secondary effects difficult. A UV effect can also be caused by photosensitizers which absorb the UV radiation; it is the photochemically generated product which damages the organisms (photodynamic reactions, see Section 6.3.4). Many organisms have repair mechanisms to remove the products of primary radiation damage to the genetic material. This also makes the identification of primary UV effects difficult.

14.1 EFFECTS OF UV RADIATION AT THE MOLECULAR LEVEL

14.1.1 Absorption and effects

Because of the aromatic π-electron system of pyrimidine and purine bases nucleic acids have a strong absorption near 260 nm. Proteins have an absorption maximum at about 280 nm due to the absorption by the aromatic amino acids phenylalanine, tyrosine and tryptophane (Fig. 14.1 and Table 14.1). Numerous lipids also absorb in the UV range due to their isolated or conjugated π-electron systems, for example the highly unsaturated fatty acids contained in complex membrane lipids. The phytohormone abscisic acid and the growth regulator indole acetic acid (IAA, auxin) also absorb in the UV

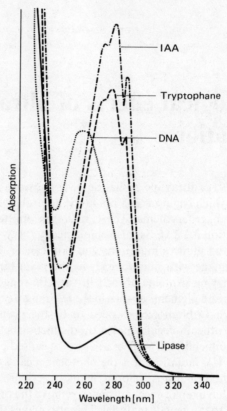

Fig. 14.1. Absorption spectra of some biologically important substances in the UV range.

range. Table 14.1 summarizes the absorption ranges of some chromophoric groups of biologically important molecules.

The primary photoreceptors can be characterized by measuring action spectra. The action spectrum for killing bacteria has a maximum near 265 nm, and strongly resembles the absorption spectrum of DNA, indicating that the UV target is located in the genome. Depending on the organism and the experimental conditions different DNA photoproducts, which cause the biological effects, can be formed by high-energy UV radiation.

14.1.2 Photoreactions of nucleic acids

The biologically most effective UV-induced photoreaction occurs at the pyrimidine base thymine. The following photoproducts can be found (Fig. 14.2):

TABLE 14.1

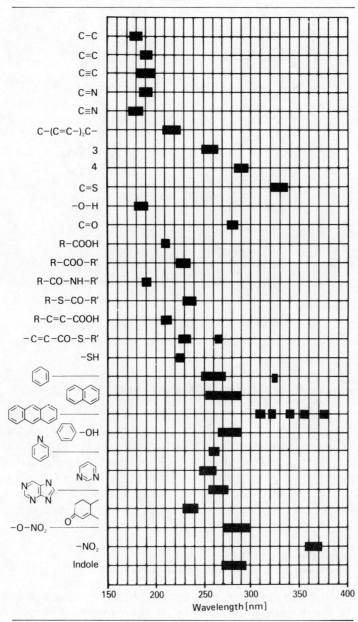

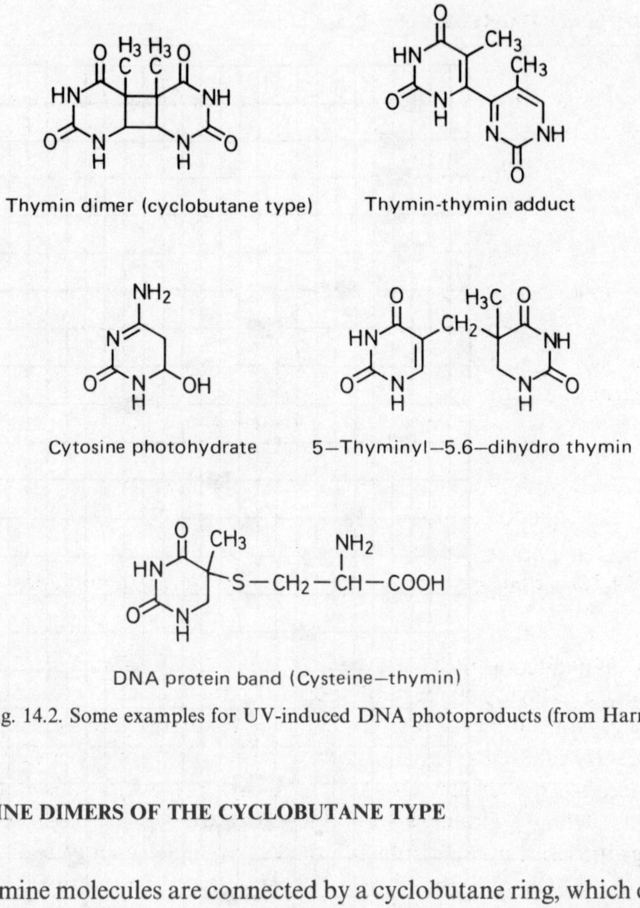

Thymin dimer (cyclobutane type)　　　　Thymin-thymin adduct

Cytosine photohydrate　　　　5—Thyminyl—5.6—dihydro thymin

DNA protein band (Cysteine—thymin)

Fig. 14.2. Some examples for UV-induced DNA photoproducts (from Harm).

1. THYMINE DIMERS OF THE CYCLOBUTANE TYPE

Two thymine molecules are connected by a cyclobutane ring, which can result in several isomers. This dimer formation occurs *in vitro* and *in vivo* with a high quantum efficiency, and may be responsible for most UV damage of the DNA. The dimerization is photoreversible (similar to the photoreversibility of the phytochrome system): UV radiation at a wavelength of 280 nm predominantly induces dimers whilst at 240 nm monomers are mainly produced.

2. PYRIMIDINE ADDUCTS

Either two thymine monomers or a thymine and a cytosine monomer form a dimer. In *Bacillus subtilis* spores we find a 5-thyminyl-5,6-dihydro-thyminyl photoproduct (Fig. 14.2). Pyrimidine adducts cannot be separated into monomers by a simple photochemical reaction like dimers of the cyclobutane type.

3. PYRIMIDINE PHOTOHYDRATES

Photohydrates are formed by addition of water to a double bond, especially in uracil and cytosine. These pyrimidine hydrates may play a role in the induction of mutations.

4. DNA PROTEIN CROSSLINKS

A DNA protein crosslink is formed after UV irradiation between a pyrimidine base and an amino acid. The bond between cysteine and thymine is especially reactive. Uracil preferentially binds to cysteine, phenylalanine and tyrosine. DNA protein crosslinks induce irreversible cell damage. In addition to pyrimidine photoproducts those of purine bases are known; their biological relevance, however, is far smaller.

14.1.3   Photoreactions in proteins, lipids and membranes

The photochemistry of proteins mainly involves the amino acids phenylalanine, tyrosine and tryptophane, with their aromatic π-electron systems, as well as histidine, cystine and cysteine. The disulfide group of cystine can be split by UV into reactive sulfhydryl groups. Since the covalent bonds of sulfur atoms are important for the tertiary structure of many proteins the effects of UV on these bonds strongly influence the structure and function of proteins. The disulfide groups can also be split indirectly by singlet energy transfer from tryptophane, which absorbs above 280 nm. Since the energy transfer to reactive centers is possible over considerable distances the longer wavelengths in the UV range are equally important.

Tryptophane can be excited either directly by UV or by energy transfer from neighboring amino acids such as phenylalanine or tyrosine. The end product is N-formyl kynurenine. After absorption of long-wavelength UV-A radiation this photoproduct can react with nucleic acids, which damages the cell or disturbs its functions. Another possible end product of irradiating tryptophane with UV is tryptamine. Photochemical reaction products of tyrosine are bityrosine and 4,4-dihydrophenylalanine (DOPA), which can also be produced in plants by a phenoloxidase. Photochemical reactions of amino acids usually deactivate the protein. In some enzymes the deactivation is due to an energy migration to the functionally important amino acids of the active center.

Lipids with isolated or conjugated double bonds can also be photochemically modified by UV absorption. Phospho- and glycolipids, which are the main components in animal and plant cell membranes, contain unsaturated fatty acids which react under UV radiation and in the presence of oxygen to lipohydroperoxides via radicals. Either radicals or singlet oxygen can be produced by photosensitization with dyes, aromatic carbohydrates or porphyrins. In the presence of protoporphyrin, a photosensitizer (precursor of hemoglobin and chlorophylls), the singlet oxygen which is produced reacts with cholesterin to form hydroperoxide, which in turn destroys the membrane fatty acids.

14.2   EFFECTS OF UV RADIATION ON CELLULAR SYSTEMS

UV-induced changes in molecular structure most probably decrease the survival probability of an organism. However, during evolution organisms have developed repair mechanisms for radiation damage, so that more UV-resistant species have evolved. The repair mechanisms do not always operate error-free, so that mutations and defects still occur in microorganisms, plants, humans and other animals.

14.2.1   Survival curves

The sensitivity of an organism to damaging radiation can be defined by its survival curve. The survival rate is usually an exponential function of radiation doses. The dose–response curves can be single-hit, shoulder and stimulation curves or biphasic curves (Fig. 14.3):

1. *Single-hit curves.* When a cell is deactivated by a single hit the survival can be defined by

$$N = N_0 \times e^{-kD}$$

   where N is the number of surviving cells, N_0 the number of initial cells and D the radiation dose; k is a species-specific constant. Due to the exponential function a semilogarithmic plot shows a linear curve (Fig. 14.3a). This type of response is usually found in viruses and enzyme deactivation curves.
2. *Shoulder curves.* At higher radiation doses the curves are also exponential. Lower doses, however, cause less damage than expected. The curve can be interpreted to represent multiple hits. The extrapolation of the exponential part indicates the number of hits (Fig. 14.3b). The shoulder may be due to a repair effect which increases the survival at lower doses and causes a deviation from a single-hit curve.
3. *Stimulation curves.* These curves show a stimulation at low doses before

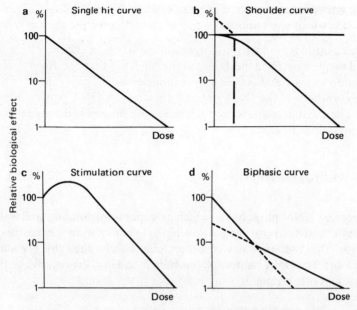

Fig. 14.3. Four typical dose–effect curves: (a) single-hit curve, (b) multiple-hit curve, (c) stimulation curve, (d) biphasic curve (from Nieman).

damage becomes manifest at higher doses. These survival curves are often found in higher plants, and are not easily explained by the hit theory (Fig. 14.3c).

4. *Bi- or multiple-phasic curves*. These curves indicate that a population consists of several subpopulations with different sensitivities. This situation can occur during aging of a population. During their exponential growth, bacteria are usually very UV-sensitive, while they are less sensitive in their stationary stage (Fig. 14.3d).

14.2.2 Repair mechanisms

The shoulder curves indicate that bacteria can possibly repair radiation damage, especially when the cells are not in the logarithmic growth phase. This can be experimentally induced by nutrition deficiency. During this resting period the repair enzymes can remove the damage to the DNA. This process is known as liquid holding recovery. There are basically three different repair mechanisms: excision repair, photoreactivation and post-replication repair.

14.2.2.1 EXCISION REPAIR

This mechanism is found in most organisms, and is independent of light. When DNA damage is detected the damaged site (such as a thymine dimer) is cleaved by an endonuclease and excised by an exonuclease. The gap is closed by a local *de novo* synthesis of the DNA strand by means of a DNA polymerase and a ligase. The excision repair seems to decrease the amount of damage that could cause mutations or cancer.

14.2.2.2 PHOTOREACTIVATION

An enzyme, DNA photolyase, which is capable of binding and splitting dimers, is activated by long-wavelength UV radiation or visible light. However, this radiation has to occur immediately after the UV damage (Fig. 14.4a). The DNA photolyase has been found in microorganisms, plants, humans and other animals.

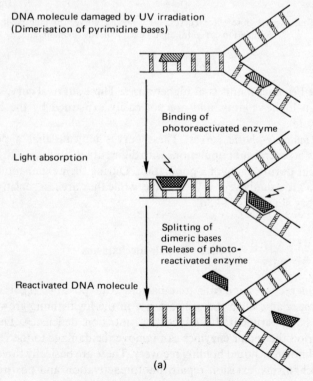

Fig. 14.4. (a) DNA repair by photoreactivation.

DNA molecule damaged by UV irradiation

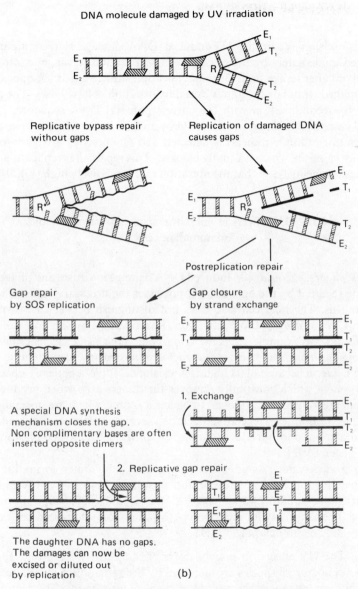

Fig. 14.4(b) Model of post-replication repair (from Beier).

14.2.2.3 POST-REPLICATION REPAIR

During excision repair the UV-induced DNA damage is recognized and removed immediately by the enzyme. Post-replication repair, in contrast, is carried out after replication. During replication either a gap is left opposite to the damaged strand or the gap is closed by insertion of false bases. The gap is closed by recombination with the correct parental DNA sequence. In the second case the false bases are removed by an SOS repair: the presence of the damage (for example dimers) is detected and functions as a trigger for the repair mechanism, which is usually blocked. This repair after replication often operates erroneously, so that the mutation rate is relatively high (Fig. 14.4b).

14.2.3   UV effects in microorganisms, invertebrates and mammalian cells

Viruses, phages, bacteria. UV-induced DNA damage in viruses and phages can often be repaired by the host cell if it possesses the necessary excision repair mechanisms. The host cell also does not distinguish between its own and foreign DNA during photoreactivation. Reactivations by the host cell have been found for example in *Escherichia coli, Salmonella typhimurium* and *Haemophilus influenza. Micrococcus radiodurans* has an exceptional position among bacteria because of its high UV resistance. It has a perfectly operating repair system which constantly removes the dimers even when produced in large numbers by high UV doses. In bacteria spores, which are normally less UV-sensitive than the vegetative bacteria, hardly any dimers are produced. Instead a spore photoproduct (see Fig. 14.2), which cannot be photoreactivated, accumulates.

Simple eukaryotes. As in bacteria, we also find dimer formation and photoreactivation in yeasts, algae and protozoa. In yeasts a UV-induced mutant (petite mutant) which has lost the capability of respiration is of interest. The sensitivity for induction of this mutation is localized both in the nucleus and the mitochondria DNA.

Insects. The UV effects on insects are interesting for two reasons. First, a number of nocturnal species are attracted by UV radiation. Second, the solar UV radiation influences the biological clock of arthropods. The removal of radiation damage by photoreactivation and other repair mechanisms has also been demonstrated in insects.

Mammalian cells. Mammalian cells such as Hela cells (cancerogenous human cells) kept in a synthetic medium have a much higher UV sensitivity than bacteria. The action spectrum indicates DNA and proteins as possible targets.

In addition to pyrimidine dimers, DNA protein crosslinks have been found. Photoreactivation has been observed in human leukocytes, but excision repair and post-replication repair mechanisms seem to dominate.

14.2.4   UV effects on vertebrates

The uncovered epidermis of many vertebrates is the main target for UV radiation. Short-term radiation can induce tanning and vitamin D synthesis. Long-term radiation, however, often induces degenerative skin diseases as well as pigmented (melanoma) and non-pigmented tumors (non-melanoma).

14.2.4.1 ERYTHEMAL ACTIVITY

The human skin consists of several layers, of which the outermost, the *stratum corneum*, has a high absorption for radiation below 300 nm. Therefore only UV radiation at longer wavelengths penetrates into the lower layers of the epidermis and the *corium*. The skin responds to increased UV-B radiation with the well-known sunburn (erythema) as the action spectrum indicates (Fig. 14.5). The visible indication of a sunburn is the red skin due to increased blood circulation in the epidermis.

Tanning is a protective mechanism of the skin against increased UV radiation. It is induced by suberythemal fluence rates and usually becomes manifest a few hours after UV application. The tanning is induced by migration of a previously synthesized chromophore melanine into the outer cell layers. A *de novo* formation of melanocytes by cell division or melanine synthesis in amelanogenous cells is also possible. A positive effect of UV radiation is the formation of the essential vitamin D_3 from 7-dehydrocholesterol (Fig. 14.6). Vitamin D_3 regulates the calcium and phosphate metabolism and prevents rachitis. UV radiation can also cause keratitis of the cornea in an unprotected eye.

14.2.4.2 MUTAGENESIS AND CANCEROGENESIS

The induction of mutations by UV radiation is based on the molecular change of DNA and erroneous repair, especially by the SOS repair. The primary UV acceptors are pyrimidine bases which, after dimer formation, induce erroneous base pair substitutions. Skin cancers also seem to be induced by photolesions. Erroneous DNA repair increases the cancer frequency as in the

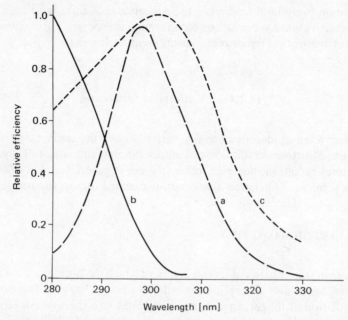

Fig. 14.5. Action spectra for (a) erythema formation in the human skin, (b) DNA damage in bacteria (from Setlow) and (c) anthocyanin synthesis in corn seedlings (from Wellmann).

case of mutagenesis. Long-term UV radiation of wavelengths between 230 nm and 320 nm can induce skin cancer.

Three types of tumors can be distinguished: epidermal tumors, subepidermal tumors and pigmented tumors (melanoma). The first two types are fairly common but relatively harmless since they do not form metastases, in contrast to the pigmented tumors. Prolonged exposition to strong solar UV radiation can be sufficient to induce skin tumors. People with fair complexions are especially sensitive, as has been demonstrated in a Scottish population which now lives in tropical Australia. Within the USA the mortality from melanomas increases from north to south, as does the exposure to UV radiation. Increased UV radiation due to possible destruction of the ozone layer by chlorfluormethanes (see Chapter 2) could induce a higher number of tumors.

14.2.5   UV effects in plants

As soon as a seedling breaks through the soil surface it is exposed to natural UV radiation. The continuous growth, however, indicates that plants can tolerate or adapt to UV radiation. When plants are grown in a greenhouse

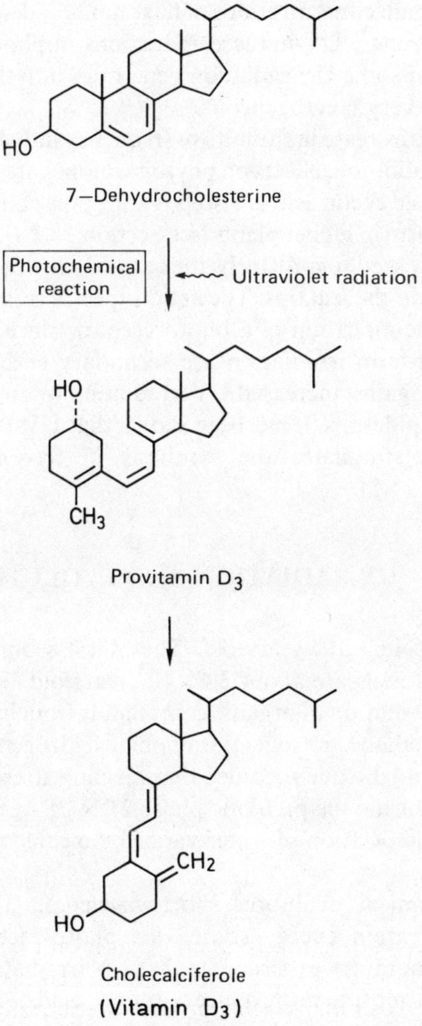

7—Dehydrocholesterine

Photochemical reaction ⟵⟍⟍ Ultraviolet radiation

Provitamin D₃

Cholecalciferole
(Vitamin D₃)

Fig. 14.6. UV-induced biosynthesis of vitamin D_3 (from Stryer).

without UV radiation they experience a UV shock when planted outside, which can kill the plant. Even plants adapted to UV radiation are impaired in their development, structure and function by increased UV radiation.

Eighty two agriculturally important species have been screened for their UV-B sensitivity: of these 24 species were susceptible, 15 sensitive and 17 highly sensitive to increased UV-B radiation, and showed a decrease in the biomass of more than 50%. Grasses were mostly UV-resistant but some species or cultivars had different sensitivities. The epidermis usually protects the inner leaf with high concentrations of flavonoids and anthocyanins, which absorb in the UV range. UV-induced reductions in photosynthetic activity show, however, that some UV radiation penetrates into the lower cell layers. The effects may be very heterogeneous.

The UV-induced decrease in the mitosis frequency in *Rumex* indicates DNA to be the target. In addition, effects on phytohormones are feasible. Photolytic degradation of indole acetic acid is also possible, especially for the blue light-induced phototropism in higher plants (see Section 13.2.1). Continuous UV-B irradiation of barley seedlings disturbs the vertical growth; this may be due to destruction of IAA in the leaf tips. The action spectra (Fig. 14.5) measured so far do not allow identification of a photoreceptor, since most of the effects detected after long-term irradiation are secondary effects. Plants possibly protect themselves against increased UV irradiation by an increased synthesis of pigments in the epidermis. It has been shown that UV-B radiation at about 300 nm effectively stimulates the synthesis of flavone glycosides and anthocyanins (Fig. 14.5).

14.3   UV RADIATION AND EVOLUTION

The earth is about 4.6×10^9 years old. The oldest sedimentary rocks with bacteria-like microfossils are about 3.4×10^9 years old. The prebiotic phase with the chemical evolution of organic compounds from inorganic gases such as water vapor, methane, nitrogen, ammonia, hydrogen sulfide as well as carbon monoxide and dioxide preceded this, forming an environment capable of producing life. During the prebiotic phase 2.2×10^9 years ago oxygen was evolved by photodissociation of water vapor by means of short-wavelength solar UV radiation.

During the chemical evolution short-wavelength UV radiation was probably also the main energy source for photoreactions. It has been demonstrated in laboratory experiments that UV irradiation of typical gases of the primeval atmosphere produces amino acids, formaldehyde and hydrogen cyanide. In aqueous solutions intensive UV radiation formed di-and tripeptides from amino acids, sugars from formaldehydes, and the purine bases adenine and guanine from hydrogen cyanide. Even nucleosides and nucleotides could be produced under UV-C irradiation from their precursors in aqueous solutions.

Principally similar photochemical reactions may have occurred during the prebiotic phase. It is not clear how self-reproducing nucleic acids have evolved from nucleotides; but it seems probable that prebiotic life has developed in the depth of the oceans protected from UV radiation. After the development of repair mechanisms primitive bacteria may have migrated into shallower waters where species developed with the capability of evolving oxygen.

The biotically produced oxygen accumulated slowly in the atmosphere, and was converted into ozone in the upper layers by solar UV radiation (see Chapter 2). The gradually forming ozone layer effectively protected life from short-wavelength UV radiation. Under these conditions life could master the transition from marine to terrestrial habitats about 400 million years ago.

14.4 BIBLIOGRAPHY

Textbooks and review articles

Caldwell, M. M. Effects of UV radiation on plants in the transition region to blue light. In: Senger, H. (ed.) *Blue Light Effects in Biological Systems*. Springer, Berlin, Heidelberg, New York and Tokyo, pp. 20–28 (1984)

Calkins, J. (ed.) *Biological Effects of Solar UV Radiation*. Plenum Press, New York (1982)

D'Ari, R. The SOS system. *Biochimie* **67**, 343–347 (1985)

Eker, A. P. M. Photorepair processes. In: Montagnoli, G. and Erlanger, B. F. (eds) *Molecular Models of Photoresponsiveness*. Plenum Press, New York, pp. 109–132 (1983)

Giese, A. C. *Living with our Sun's Ultraviolet Rays*. Plenum Press, New York (1976)

Grandolfo, M., Michaelson, S. M. and Rindi, A. (eds) *Biological Effects and Dosimetry of Nonionizing Radiation*. Vol. 49. NATO ASI Series. Series A: Life Sciences. Plenum Press, New York (1983)

Harm, W. Biological effects of ultraviolet radiation. In: Hutchinson, F., Fuller, W. and Mullins, L. J. (eds) *IUPAB Biophysics Series* 1. Cambridge University Press, Cambridge (1980)

Kiefer, J. (ed.) *Ultraviolette Strahlen*. de Gruyter, Berlin (1977)

Klein, R. M. Plants and near-ultraviolet radiation. *Bot. Rev.* **44**, 1–127 (1978)

Lake, J. A., Clark M. W., Henderson, E., Fay, S. P., Oakes, M., Scheinman, A., Thornber, J. P. and Mah, R. A. Eubacteria, halobacteria, and the origin of photosynthesis: the photocytes. *Proc. Natl. Acad. Sci.* **82**, 3716–3720 (1985)

Orton, C. G. (ed.) *Progress in Medical Radiation Physics*. Vol. 1. Plenum Press, New York and London (1982)

Parrish, J. A., Anderson, R. R., Urbach, F. and Pitts, D. *UV-A. Biological Effects of Ultraviolet Radiation with Emphasis on Human Responses to Longwave Ultraviolet*. Plenum Press, New York (1978)

Rao, K. K., Cammack, R. and Hall, D. O. Evolution of light energy conversion. In: Schleifer, K. H. and Stackebrandt, E. (eds) *Evolution of Prokaryotes*. FEMS Symp. Academic Press, London, pp. 143–173 (1985)

Further reading

Begleiter, A. and Johnston, J. B. DNA crosslinking by 3′-(3-cyano-4-morpholinyl)-3′-deaminoadriamycin in HT-29 human colon carcinoma cells *in vitro*. *Biochem. Biophys. Res. Commun.* **131**, 336–338 (1985)

Beier, W. (Hrsg.) Erzeugung, Messung und Anwendung ultravioletter Strahlen. *Fortschr. Exp. Theor. Biophys.* **25**, VEB Georg Thieme, Liepzig (1980)

Boer, J., Burger, P. M. and Simons, J. W. I. M. Interaction of far- and near-ultraviolet radiation. The occurrence of photo-augmentation and photo-recovery in cultured mammalian cells. *Mut. Res.* **125**, 283–289 (1984)

Bornman, J. F., Björn, L. O. and Akerlund, H.-E. Action spectrum for inhibition by ultraviolet radiation of photosystem II activity in spinach thylakoids. *Photobiochem. Photobiophys.* **8**, 305–313 (1984)

Caldwell, M. M. Plant response to solar ultraviolet radiation. In: Lange, O. L., Nobel, P. S., Osmond, C. B. and Ziegler, H. (eds) *Encyclopedia of Plant Physiology*, new series, Vol. 12A: *Physiological Plant Ecology*. Springer, Berlin (1981)

Canuto, V. M., Levine, J. S., Augustsson, T. R., Imhoff, C. L. and Giampapa, M. S. The young sun and the atmosphere and photochemistry of the early earth. *Nature* **305**, 281–286 (1983)

Chavez, E. and Cuellar, A. Inactivation of mitochondrial ATPase by ultraviolet light. *Arch. Biochem. Biophys.* **230**, 511–516 (1984)

Cunningham, M. L., Johnson, J. S., Giovanazzi, S. M. and Peak, M. J. Photosensitized production of superoxide anion by monochromatic (290–405 nm) ultraviolet irradiation of NADH and NADPH coenzymes. *Photochem. Photobiol.* **42**, 125–128 (1985)

Diffey, B. L., Whillock, M. J. and McKinlay, A. F. A preliminary study on photoaddition and erythema due to UVB radiation. *Phys. Med. Biol.* **29**, 419–425 (1984)

Döhler, G. Effect of UV-B radiation on biomass production, pigmentation and protein content of marine diatoms. *Z. Naturf.* **39c**, 634–638 (1984)

Döhler, G. Effect of UV-B radiation (290–320 nm) on the nitrogen metabolism of several marine diatoms. *J. Plant Physiol.* **118**, 391–400 (1985)

Fraikin, G. Y., Strakhovskaya, M. G. and Rubin, L. B. Photomimetic effect of serotonin on yeast cells irradiated by far-UV radiation. *Photochem. Photobiol.* **35**, 799–802 (1982)

Graem, N. and Povlsen, C. O. Acute effects of ultraviolet light B on morphology and epidermal cell kinetics in human skin transplanted to nude mice. *Exp. Cell Biol.* **52**, 311–319 (1984)

Häder, D.-P. Effects of UV-B on motility and photoorientation in the cyanobacterium, *Phormidium uncinatum. Arch. Microbiol.* **140**, 34–39 (1984)

Häder, D.-P. Effects of UV-B on motility and photobehavior in the green flagellate, *Euglena gracilis. Arch. Microbiol.* **141**, 159–163 (1985)

Ikai, K., Tano, K., Ohnishi, T. and Nozu, K. Repair of UV-irradiated plasmid DNA in excision repair deficient mutants of *Saccharomyces cerevisiae. Photochem. Photobiol.* **42**, 179–181 (1985)

Imray, P., Mangan, T., Saul, A. and Kidson, C. Effects of ultraviolet irradiation on the cell cycle in normal and UV-sensitive cell lines with references to the nature of the defect in xeroderma pigmentosum variant. *Mutation Res.* **112**, 301–309 (1983)

Iwanzik, W., Tevini, M., Dohnt, G., Voss, M., Weiss, W., Gräber, P. and Renger, G. Action of UV-B radiation on photosynthetic primary reactions in spinach chloroplasts. *Physiol. Plant.* **58**, 401–407 (1983)

Larcom, L. L. and Rains, C. A. Far UV irradiation of DNA in the presence of proteins, amino acids or peptides. *Photochem. Photobiol.* **42**, 113–120 (1985)

Mathews-Roth, M. M. and Krinsky, N. I. Carotenoid dose level and protection against UV-B induced skin tumors. *Photochem. Photobiol.* **42**, 35–38 (1985)

Mirecki, R. M. and Teramura, A. H. Effects of ultraviolet-B irradiance on soybean. V. The dependence of plant sensitivity on the photosynthetic photon flux density during and after leaf expansion. *Plant Physiol.* **74**, 475–480 (1984)

Murali, N. S. and Teramura, A. H. Effects of ultraviolet-B irradiance on soybean. VI. Influence of phosphorus nutrition on growth and flavonoid content. *Physiol. Plant.* **63**, 413–416 (1985)

Murphy, T. M. Membranes as targets of ultraviolet radiation. *Physiol. Plant.* **58**, 381–388 (1983)

Murphy, T. M., Hurrell, H. C. and Sasaki, T. L. Wavelength dependence of ultraviolet radiation-induced mortality and K^+ efflux in cultured cells of *Rosa damascena. Photochem. Photobiol.* **42**, 281–286 (1985)

Niemann, E.-G. Strahlenbiophysik. In: Hoppe, W., Lohmann, W., Markl, H. and Ziegler, H. (eds) *Biophysik.* Springer, Berlin, pp. 300–312 (1982)

Okaichi, K., Tano, K., Ohnishi, T. and Nozu, K. Removal of pyrimidine dimers in UV-irradiated spores of *Dictyostelium discoideum* during germination. *Photochem. Photobiol.* **41**, 649–653 (1985)

Park, Y. K., Gange, R. W., Levins, P. C. and Parrish, J. A. Low and moderate irradiances of UVB and UVC irradiation are equally erythemogenic in human skin. *Photochem. Photobiol.* **40**, 667–669 (1984)

Parson, P. G. and Hayward, I. P. Inhibition of DNA repair synthesis by sunlight. *Photochem. Photobiol.* **42**, 287–293 (1985)

Peak, J. G., Peak, M. J., Sikorski, R. S. and Jones, C. A. Induction of DNA-protein crosslinks in human cells by ultraviolet and visible radiations: Action spectrum. *Photochem. Photobiol.* **41**, 295–302 (1985)

Peak, M. J., Peak, J. G. and Jones, C. A. Different (direct and indirect) mechanisms for the induction of DNA-protein crosslinks in human cells by far- and near-ultraviolet radiations (290 and 405 nm). *Photochem. Photobiol.* **42**, 141–146 (1985)

Quarless, S. A. and Cantor, C. R. Analysis of RNA structure by ultraviolet crosslinking and denaturation gel electrophoresis. *Analyt. Biochem.* **147**, 296–300 (1985)

Senter, P. D., Tansey, M. J., Lambert, J. M. and Blättler, W. A. Novel photocleavable protein crosslinking reagents and their use in the preparation of antibody-toxin conjugates. *Photochem. Photobiol.* **42**, 231–237 (1985)

Setlow, R. B. The wavelength in sunlight effective in producing skin cancer: theoretical analysis. *Proc. Natl Acad. Sci.* **71**, 3363–3369 (1974)

Shetlar, M. D. Cross-linking of proteins to nucleic acids by ultraviolet light. *Photochem. Photobiol. Rev.* **5**, 107–197 (1980)

Shiroya, T., McElroy, D. E. and Sutherland, B. M. An action spectrum of photoreactivating enzyme from sea urchin eggs. *Photochem. Photobiol.* **40**, 749–751 (1984)

Takahama, U., Egashira, T. and Nakamura, K. Photoinactivation of a *Chlamydomonas* mutant (NL-11) in the presence of methionine: Roles of H_2O_2 and O_2. *Photochem. Photobiol.* **41**, 149–152 (1985)

Teramura, A. H. Effects of ultraviolet-B radiation on the growth and yield of crop plants. *Physiol. Plant.* **58**, 415–427 (1983)

Teramura, A. H., Forseth, I. N. and Lydon, J. Effects of ultraviolet-B radiation on plants during mild water stress. IV. The insensitivity of soybean internal water relations to ultraviolet-B radiation. *Physiol. Plant.* **62**, 384–389 (1984)

Teramura, A. H., Perry, M. C., Lydon, J., McIntosh, M. S. and Summers, E. G. Effects of ultraviolet-B radiation on plants during mild water stress. III. Effects on photosynthetic recovery and growth in soybean. *Physiol. Plant.* **60**, 484–492 (1984)

Tevini, M., Iwanzik, W. and Thoma, U. Some effects of enhanced UV-B radiation on the growth and composition of plants. *Planta* **153**, 388–394 (1981)

Tevini, M. and Pfister, K. Inhibition of photosystem II by UV-B radiation. *Z. Naturf.* **40c**, 129–133 (1985)

Tevini, M., Thoma, U. and Iwanzik, W. Effects of enhanced UV-B radiation on germination, seedling growth, leaf anatomy and pigments of some crop plants. *Z. Pflanzenphysiol.* **109**, 435–448 (1983)

Umlas, M. E., Franklin, W. A., Chau, G. L. and Haseltine, W. A. Ultraviolet light irradiation of defined-sequence DNA under conditions of chemical photosensitization. *Photochem. Photobiol.* **42**, 265–273 (1985)

Wang, S. Y. (ed.) *Photochemistry and Photobiology of Nucleic Acids.* Academic Press, New York (1976)

Webb, R. B. Lethal and mutagenic effects of near ultraviolet radiation. *Photochem. Photobiol. Rev.* **2**, 169–261 (1977)

Wellmann, E. UV radiation in photomorphogenesis. In: Shropshire, W. and Mohr, H. (eds) *Encyclopedia of Plant Physiology*, new series, Vol. 16B: *Photomorphogenesis*. Springer, Berlin (1983)

Wellmann, E., Schneider-Ziebert, U. and Beggs, C. J. UV-B inhibition of phytochrome-mediated anthocyanin formation in *Sinapis alba* L. cotyledons. Action spectrum and the role of photoreactivation. *Plant Physiol.* **75**, 997–1000 (1984)

Wilkins, R. J. Photoreactivation of UV damage in *Sminthopsis crassicaudata. Mutation Res.* **111**, 263–276 (1983)

World Health Organisation, *Ultraviolet radiation. Environmental Health Criteria* **14**, Kirjapaino, Vammala (1979)

Young, A. R., Guy, R. H. and Maibach, H. I. Laser doppler velocimetry to quantify UV-B induced increase in human skin blood flow. *Photochem. Photobiol.* **42**, 385–390 (1985)

Index

311